RESURRECTION SCIENCE

CONSERVATION, DE-EXTINCTION
AND THE PRECARIOUS
FUTURE OF WILD THINGS

M. R. O'CONNOR

St. Martin's Press
New York

Special appreciation to the Alfred P. Sloan Foundation's Program for the Public Understanding of Science, Technology, & Economics for their financial support of the writing of this manuscript.

Grateful acknowledgement is made to the editors of the periodicals in whose pages excerpts of this book first appeared:

Passages of Chapter 1 appeared in Guernica.
Excerpts of Chapter 6 appeared in The Atlantic.

www.stmartins.com

Library of Congress Cataloging-in-Publication Data

O'Connor, M. R., 1982-
   Resurrection science : conservation, de-extinction and the precarious future of wild things / M.R. O'Connor.
      pages   cm
   Includes bibliographical references.
   ISBN 978-1-137-27929-3
   1. Animal diversity conservation—Anecdotes. 2. Animal diversity conservation—Moral and ethical aspects. 3. Endangered species—Anecdotes. 4. Wildlife conservationists—Anecdotes. I. Title.
   QL82.O26   2015
   591.68--dc23

                                                                              2015004485

ISBN 978-1-137-27929-3 (hardcover)
ISBN 978-1-4668-7932-4 (e-book)

Our books may be purchased in bulk for educational, business, or promotional use. Please contact your local bookseller or the Macmillan Corporate and Premium Sales Department at 1-800-221-7945, extension 5442, or by email at MacmillanSpecialMarkets@macmillan.com.

Design by Letra Libre Inc.

First Edition: September 2015

10   9   8   7   6   5   4   3   2   1

For Joaquín, a wild one

"Extinction is truly forever; once a group dies out, the hundred thousand unpredictable stages that led to its origin will never be repeated in precisely the same way."

—Stephen J. Gould, *An Urchin in the Storm* (W. W. Norton)

"One morphologically normal bucardo female was obtained by caesarean section. The newborn died some minutes after birth due to physical defects in lungs. Nuclear DNA confirmed that the clone was genetically identical to the bucardo's donor cells."

—"First birth of an animal from an extinct subspecies (Capra pyrenaica pyrenaica) by cloning." *Theriogenology,* January 23, 2009

CONTENTS

INTRODUCTION

When I was a kid in the 1990s, the world often felt like it was on the verge of catastrophe. At my California public school, to offset drought, teachers taught us to conserve water while brushing our teeth. I saw news stories about environmental crises such as burning rainforests and acid rain. None of these issues loomed as large in my imagination, though, as the idea that the world is in the midst of a "Sixth Extinction." In the early 1990s, Richard Leakey, the famous Kenyan paleoanthropologist, used this term to describe the phenomenon of disappearing species; and it gained wide acceptance in the public domain, giving gravitas and urgency to media reports and conservation campaigns. When I was in middle school, I heard predictions like those made by the British environmentalist Norman Myers, who estimated that 50 percent of all species at that time would go extinct in the twenty-first century. Harvard biologist Edward O. Wilson estimated as many as 27,000 species were going extinct every year. Trying to understand these numbers is bewildering for anyone. As a kid, the idea that every hour three species disappeared was incomprehensible. I didn't understand how there could be so many species on earth that we could lose so many. But I internalized these statistics, and developed a concern for the fate of species rooted in two beliefs: extinction is bad, and saving species from extinction is good.

Then a few years ago, I found myself in the back of the Bronx Zoo's Reptile House peering through a little pane of glass into a room full of terrariums. Through the condensation on their walls, I barely made out dozens of blurry, yellow frogs climbing over green moss. I wanted to get a closer look but the room was off-limits: the only people allowed inside were the herpetologists responsible for the frogs' care, and even they had to disinfect the bottom of their shoes with bleach. The frogs inside this bio-secure room were incredibly rare, one of two populations left in the world, both in captivity. The waterfall where they had come from in the Tanzanian rainforest was now the site of a hydroelectric dam, and the species was confined to these terrariums, painstakingly kept hydrated with an artificial misting system and fed specially bred insects. It was like peering into a hospital ward at a patient on life support.

The great lengths that had been taken to save the frogs from extinction intrigued me, and a year later I was in Tanzania interviewing some of the central characters of the Kihansi spray toad story. I expected to learn a lot about conservation biology. Instead I found myself on a crash course in national politics, development economics, racial privilege, bureaucratic subterfuge, and environmental ethics. I realized that my belief that conservation is self-evidently good was actually a social and cultural bias. By the time I made it into the country's remote forest where the frogs once thrived, I was having thoughts that would have previously struck me as unenlightened: maybe they should have just let those toads disappear. Staring at their former habitat, a mere two hectares of wetlands, the species struck me as a kind of evolutionary whimsy, extraordinary for its perfect adaption to a waterfall and beautiful for its incredible rarity. But now they were confined to a little bathysphere suspended in a world aflood in disaster. Was that better than extinction? I couldn't say for sure. The millions of dollars that had been spent on their conservation seemed almost cruel in the context of the immense poverty of rural East Africa. Saving species, it turned out, was not a simple thing with bad guys and heroes and tidy endings.

In the years following my reporting in Tanzania, I began looking at other cases of threatened species and the efforts to protect them. Tragically, there are thousands of species to choose from, each conservation effort as scientifically fascinating and ethically complex as the next. My goal hasn't been to write comprehensively about the field of conservation; instead I've focused on dramatic cases of animals that are on the precipice of dying out or have already vanished. The extreme nature of these stories crystalizes the questions at the heart of our evolving morals and relationship to the natural world. How can humans coexist with species in the modern world when our existence and their survival so often appear pitted against one another? What should we preserve of wilderness as we head toward a future of incredible technological control over biology? Is nature here to serve our interests, or is its independence worthy of protection? I discovered that contrary to a common perception that much of biology has been explained, scientists are making new and incredible discoveries today that give us a glimpse into the complex relationship between genes, ecology, and evolution. And at a time of rapid environmental change on earth—as industrialization, globalization, and human sprawl go unchecked—these discoveries are much more than intellectual marvels; they present clues as to how we might prevent snuffing out the existence of other species.

Until very recently, humans never really cared much about whether species disappeared. History is littered with our indifference—dodo birds, great auks, 24-rayed sunstars, dusky seaside sparrows, Bernard's wolves. For most of human civilization, no one really believed species *could* go extinct. The environmental ethic planted in me as a kid probably wouldn't have made sense even a hundred years ago to most people, who generally believed the earth's bounty was intended for the benefit of human survival. It wasn't until the early twentieth century—following the work of individuals such as John Muir, Henry David Thoreau, and Aldo Leopold—that an ethic in which species were valued was born in the modern mind. When the environmental movement gained traction in the 1960s, the linchpin in the argument to save species was the threat of their extinction. Extinction, it has been said, is the middle name of conservation biology, a discipline that coalesced in the late 1970s at a time when the impacts of humanity's disturbance of the earth's ecosystems were manifesting themselves and documented by scientists. This era has been described as a new geological epoch in which humans are a force of nature. Called the Anthropocene, its Noahs are conservation biologists, people who dedicate their professional lives to saving species.

The field of conservation biology is a crisis discipline; Michael Soulé, one of its earliest visionaries, wrote that its relationship to the biological sciences is analogous to that of surgery to physiology and war to political science. Facing a Sixth Extinction, it's no surprise perhaps that conservation biologists can be a gloomy bunch. Conservationists themselves have said that the field breeds a culture of despair. And at times, their pessimism threatens to undermine the cause. "A society that is habituated to the urgency of environmental destruction by a constant stream of dire messages from scientists and the media will require bigger and bigger hits of catastrophe to be spurred to action," wrote biologists Ronald Swaisgood and James Sheppard in 2010.

The fact is, few of the dire predictions I heard as a kid about the Sixth Extinction have come to pass. It's generally thought that individual species last on average about a million years, and the idea that we are in the midst of a sixth mass extinction is based on an estimate that extinction rates have increased above this background rate. In 2000, the United Nations' Millennium Ecosystem Assessment estimated that extinction rates are now as much as 1,000 times this "normal" background rate and could increase to 10,000. But in the last 500 years, the number of species that have completely disappeared, what is known as a "true extinction," is less than 900. Some analyses show the rate of extinctions among birds and mammals has actually decreased since a peak in the eighteenth and nineteenth centuries; in 1900, they were going extinct at a rate of 1.6 per year, but that number has dropped to 0.2.

What's really going on? If we are in the midst of the Sixth Extinction, why aren't more species disappearing? Scientists know about this discrepancy between estimated extinction rates and actual extinctions and have chalked it up to something called the "extinction debt," the idea that species can be "committed" to extinction because of the loss of their habitat or shrinking population size, but can take a while to actually go extinct. But a few years ago, two researchers, Fangliang He and Stephen Hubbell, realized the discrepancy was in part due to flawed math. The common formula used to estimate species loss as it relates to habitat destruction can inflate extinction rates by as much as 160 percent. When He and Hubbell published their findings in 2011 in *Nature*, it was tricky news to deliver and created much controversy.

A few of years later, the journal *Science* published a paper whose authors further challenged the perception that untold numbers of species are disappearing before they can be discovered or named, a concern partly based on an oft-cited statistic that there are as many as 30 million to 100 million species on earth. Instead, the authors argued that there are in all likelihood around 5 million species on earth. "A meeting of conservation biologists or ecologists is hardly complete without worries about extinction rates, that many millions of species are yet to be discovered, and that the taxonomic workforce is decreasing," wrote the authors. "We do not dispute that we are in a human-caused mass extinction phase with many species committed to extinction, but actual extinctions have been fewer than arguably expected. With a realistic surge effort, most species could be named within the present century."

Undoubtedly, it's good news that current extinction rates have largely been overestimated. But the fact that extinctions aren't occurring at the rate previously believed can misconstrue the problem of habitat loss. Wild, untrammeled places are disappearing, and, with them, wild creatures independent from human influence. In 2009 the European Commission and World Bank published a study that showed a mere 10 percent of the earth's land today qualifies as "remote," meaning it takes more than forty-eight hours to travel to it from a city. As humans and their activities have spread across the globe to extract resources, plant crops, and build cities and roads, thousands of species inhabit slivers of their previous ranges, surviving in isolated populations with few places to expand and faced with a loss of genetic fitness and an increased vulnerability to climatic shifts, disease, and natural disasters. Tigers, for example, inhabit less than 7 percent of the space they inhabited a century ago. Caribou have lost half of their historic range in the last 100 years. The World Wildlife Fund (WWF) estimates that, on average, vertebrate species populations have shrunk by half since 1990. Anthropogenic global warming has exacerbated the problem of disappearing habitat and shrinking abundance. Few landscapes around the world remain untouched by climate change today. And as they have for millennia, changes in climate are acting as selective pressures on species. For those animals that can't withstand changes to their environments, migrate or adapt fast enough, their survival often depends on human intervention. An estimated 4,000 to 6,000 vertebrate species will need captive breeding over the next 200 years in order to mitigate extinction threats. Faced with these emergencies, the urgency to *do something* to save these species seems like a good ethical argument for action. But what actions we take have enormous consequences for the evolution of species.

When Darwin published *On the Origin of Species* in 1859, he believed that evolution through natural selection was a gradual process that took place over millions of years. But species such as the White Sands pupfish of New Mexico, which I write about in this book, have shown in recent years that natural selection can occur at rapid speeds, over mere decades. What this means is that humans are in the midst of an unplanned experiment of influencing the evolution of the planet's biodiversity. The same forces driving extinctions—anthropogenic global warming, degraded habitats, overexploitation, disease, invasive species—are shaping the evolutionary trajectories of species. And which animals we prioritize, and how we choose to save them, tinkers with the biosphere as a whole.

The predominant thinking in conservation biology has been that preventing extinctions and hoping species will return to a fully conserved state—meaning free of direct human management—is enough. In an era of climate change, conservationists are realizing this is no longer a realistic expectation. Returning to a prelapsarian state of untouched wilderness, if one ever truly existed, is impossible. "It's hard to future gaze to the next few hundred years, but unfortunately I think that's what is going to happen: life on this planet will likely be managed," Brad White, a conservation geneticist, told me. "It might not be as totally managed as a zoo or farm animals, but we are getting to that point." The evolutionary ecologist Michael

Kinnison described it this way to me: "Early on, the goal was to save the organisms from the environment. Bring them into captivity, and if you treated them nicely and not too biased in how you bred them, it won't be an issue. Now there is much more of an understanding that organisms will adapt to those environments. In the process of trying to save them, we change them." The irony of this age is that often the more we intervene to save species, the less "wild" and autonomous they become.

The looming ethical question is now whether or not humans, recognizing their evolutionary impact on species, should begin to consciously direct or engineer evolution in the direction they want it to go. Sometimes called prescriptive evolution or directed evolution, it might take the form of imbuing a species with characteristics that can help it survive environmental impacts down the road; translocating animals; or creating new, more resilient hybrids. Engineering biological processes in this way represents a kind of devil's bargain for conservationists, who have traditionally separated people and nature. "When you are talking about messing with evolution, you are talking about the heart and core of what is special about this planet," said Scott Carroll, a biologist and founder of the Institute for Contemporary Evolution, in an interview. But Carroll, a leader in the nascent field of applied evolutionary biology, responds to skeptics of prescriptive evolution by pointing out that we are doing it anyway. "It's happening in this unplanned, unaware way in every living and breathing moment. We're proposing a more thoughtful way. I don't think we can come up with a sustainable relationship to the planet if we don't become cognizant evolutionary organisms."

The most explicit form of human bioengineering may be de-extinction, the ability to bring species that have already disappeared back to life with the goal of one day reintroducing them to their historical range. The technology of "resurrection science" is real and upon us. Scientists have not only successfully cloned endangered animals such as the European mouflon and the African wildcat, they are also working to bring back animals that are already extinct. In 2009, Spanish scientists successfully resurrected the Pyrenean ibex in the womb of a surrogate, though the animal lived just a few minutes after birth. International efforts are underway to bring back mammoths. These attempts to repopulate the modern landscape with extinct fauna rest on an intriguing ethical argument: that humans have a moral responsibility to make amends for overexploitation by our ancient and recent ancestors.

Take the case of the passenger pigeon, whose potential de-extinction has become symbolic of both our incredible faith in science to solve our ecological problems as well as a metaphysical predicament. Is a bird born of human ingenuity in the laboratory the same as a bird born of natural selection in the wild? Or is it a case of what sociologists call bio-objectification, defined as the process by which life is made an object by humans? In 1982, Robert Elliot penned a paper called "Faking Nature" that rebuked the idea that an ecosystem disturbed or damaged by humans could be restored to its original state or has equal value to wilderness. Nature, wrote Elliot, is "not replaceable without depreciation in one aspect of its value which has to do with its genesis, its history." It seems that today we have to decide whether that genesis in the wild is something we value.

Some scientists see de-extinction as irrelevant to the real grunt work of fighting for the survival of species. "For people who are doing this work, the passenger pigeon stuff is just an offensive conversation," one biologist told me. "It's publicity for newspaper articles." There is real concern that the very idea that de-extinction is possible will weaken the will of the public and policy makers to protect endangered species or habitat.

I found the individuals working on de-extinction projects to be brilliant, and a few downright inspiring. But not many have shown how we will put resurrected animals back into the world at a time when humans can barely coexist with extant species. Florida panthers were once thought to be extinct in the mid-twentieth century. By the time a remnant population was discovered by a legendary predator hunter in southern Florida, the animals had characteristics of severe inbreeding. After a genetic rescue operation was carried out in the early 1990s, the number of Florida panthers increased, but the animals today are limited to a paltry fraction of their former habitat, bordered on all sides by Florida's booming population. "It's a success story because the panthers are sturdy, but they are raising them in a cage," Rocky McBride, someone who has tracked the panthers and other predator species for decades, told me.

After immersing myself in stories about extinction, the term *Sixth Extinction* has begun to feel unhelpful for grasping the scale and nature of the problem of diminishing biodiversity today. It is a monolithic idea; many of us are conscious that something terrible is happening to earth's creatures even as the complexity of the problem eludes our full comprehension. Indeed, the idea of mass extinction can be so overwhelming, eliciting feelings of guilt and fear, that it eventually becomes an impotent fact in the same way that a million deaths is a statistic rather than a tragedy. In the stories that follow, I've tried to give flesh to a phenomenon that haunts the periphery of our awareness but is rarely seen or experienced directly. These stories are about just a handful of the animals that are on life support today, others that have already disappeared, and the people who discover, study, track, hunt, love, obsess, philosophize, save, and try to resurrect them.

1

AN ARK OF TOADS
Nectophrynoides asperginis

On a blazing hot afternoon, Kim Howell sat in his office at the University of Dar es Salaam, crammed with the detritus of forty years of biological research, and plucked a small glass jar from dozens of bottles balanced on a shelf.

"This is it," he said. "It really doesn't look like much."

Floating in the faintly amber liquid was a tiny frog. Brownish skin, pointy nose, it belied nothing significant in appearance. Howell, a kindly white-haired giant with Coke-bottle glasses, had other jars that looked more interesting—floating bats and snakes, each one the subject of his wide-ranging biological curiosity. But perhaps none was so precious as the tiny frog, a species listed on the Convention on International Trade in Endangered Species (CITES) most-restricted list, Appendix I, reserved for rare and critically threatened species of the world, such as rhinoceroses and tigers. He was the first person in the world to discover the tiny amphibian and gave the species its name, *Nectophrynoides asperginis*, inspired by the Latin *aspergo*, meaning "spray."

It certainly wasn't the first species Howell discovered. "I found new species of spider, tapeworm, I've had stuff named after me," he said. Among them are a shrew and a subspecies of bird. "What else?" he wonders aloud, trying to reach back through the decades. "A lizard. I think the bird is called a yellow streaked green bull. And then the lizard is called *Lygodactylus kimhowelli*." What is it like to discover a new species? I asked. "It is exciting when you see something that's new. You don't want to say you're the *first* person to have seen it before, but nobody has ever described it or photographed it or bothered to say, 'Yes, this one is probably new.'" Nonetheless, he pointed out, the novelty can wear off. "It's fairly normal for a biologist who's working with smaller animals to find new species. If you are an insect person you can find hundreds. Or mites or ticks. If you work on elephants and buffalo the chances are much smaller of course."

Howell's office at the University of Dar es Salaam feels like a universe away from Pittsfield, Massachusetts, where he was born and raised. His ticket out of the small industrial town was an acceptance letter to Cornell University, where he paid for his degree in vertebrate zoology by working in the school's Laboratory of Natural Sound, preserving archival recordings of birdcalls collected in Africa during the early twentieth century. After four years, Howell wanted nothing more than to go to Africa himself, though the Vietnam War also played a role. As a conscientious objector, he needed an alternative service approved by the American government. In 1969 he chose a "wild card option in the middle of nowhere Zambia" where he taught science at a remote elementary school. At the end of that first year, he traveled north to Tanzania where he worked at a school for children of South Africa's apartheid refugees before deciding to stay for good. Howell has lived in Tanzania ever since, raising a daughter with his wife and teaching at the university.

In the early 1990s, Howell was looking through a local newspaper and took note of an unusual ad for employment. Placed by an agency called Norconsult, a Norwegian engineering firm, the ad was soliciting environmental consultants. "There was this hydropower project that was going to be done, and they were looking for someone to look at birds," said Howell.

"The location was so far away, I'd never been there. I didn't even know where it was." Howell decided to write to the company but didn't hear anything back for nearly two years. Then, seemingly out of the blue, a man walked into Howell's office and asked if he was interested in doing some studies related to a hydropower dam in the Udzungwa Mountains, one of the south-ernmost areas of the Eastern Arc Mountains. "I said, 'Sure.' How often do you go someplace no one's ever been to before and get paid for it?" recalled Howell.

In those days the journey from Dar es Salaam to the Udzungwa Moun-tains took a full day on a dirt road that roughly paralleled the rail lines of the Tanzanian-Zambian railroad, an early development project by the Chinese in Africa laid down in 1968. Villagers in the region use the train tracks as a footpath through the banana trees, sugarcane fields, and lush floodplains of the Kilombero Valley. The Eastern Arc Mountains are made up of basement rock from the Precambrian eon, some of it dating back 3.2 billion years. Around 30 million years ago, the crust fissured, cracked, and faulted, and pushed the rock into the form of a crescent-shaped mountain range cam-bering through East Africa. The uplift separated the Arcs from the main Guineo-Congolian forest of west and central Africa, birthing a kind of ar-chipelago of primeval forest that was kept stable by consistent temperatures and high rainfall from the nearby Indian Ocean.

The mountain range is sometimes called Africa's "Galapagos Islands" because there are thirteen mountain "islands," each with their own unique variations of species and habitat but part of the same original geological event and climate. Each of these islands became a laboratory of natural se-lection by virtue of its isolation, giving rise to unique trajectories of species and an endemism unrivaled in the world. Biologists today have recorded ninety-six vertebrates and over 800 endemic plant species (including thirty-one species of African violet alone) in the Arcs. The stability of the climate may also have reduced the rate of extinction, which scientists determine by the number of genetically ancient species they find present in the for-ests. DNA analysis of some forest birds in the Eastern Arcs shows lineages stretching back to the early Miocene epoch some 20 million years ago. Much of the fauna reveals a greater connection to Madagascar than continental Africa, with other birds showing commonality with subspecies originating in Southeast Asia—back when a single continent called Pangaea covered the globe.

Deep in the Udzungwa Mountains, a river of water cut through the forest and over a steep gorge, creating a plunging waterfall. From top to bottom, the gorge is roughly two miles long and drops nearly 3,000 feet. Unlike nearly every other river and stream in Tanzania, what became known as the Kihansi River didn't shrink during the dry season. Twelve months of the year, the waterfall inside the gorge was so powerful that the cascades could be seen from miles away against the thick verdant rainforest— majestic and inaccessible. Around 1984, the Tanzanian government began studying the waterfall as a possible site for a hydropower project, recognizing that it appeared to have the perfect combination of features needed to generate an incredible amount of electricity in a country that had very little. "There are two major factors that play an important role," said Rafik Hajiri, a water resources expert. "One is the stable hydrology and the second is the drop. And Kihansi has both. It has one of the most stable hydrologies that we know of in the country. And second, the drop is just amazing."

It was the Tanzanian government's hydropower project—funded by the World Bank—that Howell and a team of biologists were hired to conduct an environmental impact assessment (EIA) for. But from the moment they arrived they knew something wasn't right. "We were conducting our studies literally one step ahead of the bulldozers," recalled John Gerstle, the man who walked into Howell's office in 1994 and ran the assessment for Norconsult. Normally, Gerstle explained, ground for such a large-scale development project wouldn't be broken until the EIA was complete. In fact, the World Bank had commissioned an EIA in 1991, but it had later been determined to be completely inadequate.

A slim, fifty-page document, it was written by a Kenyan PhD student who undertook two ten-day trips to the area around Kihansi, interviewing villagers and showing them pictures in field guides of birds and mammals to see which fauna were present in the forest. At the end of the report, the student surmised: "The area to be lost is so small that it will not be a seri-ous environmental loss and is most unlikely to threaten extinction of any of the endemic species as they also exist in other parts of the U[d]zungwa forest. The loss of habitat is a small price to pay for the economic benefits of power generation." Anna Maembe, a senior staff member at Tanzania's National Environmental Management Council, explained that at the time, the country had no legal requirements regarding EIAs. "People did it to go to the banks to get loans and so on," she said. "It wasn't based on government support, on a legal premise."

The World Bank did have internal policies for undertaking EIAs and had rated the Kihansi project "Category A," requiring a full assessment. "There was an attempt to do that [assessment], but the problem was it wasn't done as comprehensively as one would have expected," said Jane Kibbassa, senior environmental specialist at the World Bank in Tanzania. In 1994, the decision to finance the hydropower project came before the World Bank's board of directors, and it was in part the EIA that they considered in their decision to approve the $200 million loan to the government of Tanzania to start construction on the project. When the European Bank, Norway, Sweden, and Germany's development agencies came on board as donors a year later, however, they balked at the assessment and made a new one a condition of their participation. "It was a bit late," said Gerstle, "because the decision had already been made to build the project. The reasoning was they were just desperate to get new capacity because there were continuous rolling blackouts. It was very severe."

Tanzania was, and remains, a country mostly in the dark. Shopping at a mall in Dar es Salaam, a sudden blackout will leave people standing in a department store checkout line in pitch-blackness until a generator hopefully kicks in. The power outages are not just a frequent annoyance: electricity shuts off without notice multiple times a week, and outages can last for hours, grinding to a halt the activities of businesses and schools. For three months in 2009, the semi-autonomous archipelago of Zanzibar had zero electricity after an old, ill-maintained undersea cable that con-nected it to the power grid on the Tanzanian mainland failed. Villagers in

Zanzibar recall their childhood as times when they had more consistent and cheaper access to electricity (and hence running water and refrigeration) than they do today. The problem is enormous in scale: access to electricity in Tanzania, where 73 percent of the population survives on less than $2 per day, is around 39 percent for people in urban areas, but only 2 percent of people in rural areas can get power, according to the United Nations Development Program. Tanzania's per capita use of electricity is small even by sub-Saharan levels. The people of the Democratic Republic of Congo, which has been roiled in civil war for decades, have greater access to electricity than Tanzanians. North Korea produces more electricity.

"Tanzania is seriously underinvested and has been for a long time in modern energy," said John McIntire, former World Bank country director for Tanzania, Uganda, and Burundi. "What this means is that people don't have access to the cold chain for preserving products, for keeping medicines cold. And you get other indirect effects of a lack of electricity. It's a thing that allows people to work at night, which is important in a hot climate. It's a labor-saving device for some things that you just can't do by applying more and more labor." But the conflict between protecting biodiversity and eradicating poverty is uncomfortably direct. "The international community has to understand that these countries need electricity," argued McIntire. "It's not just for the rich countries to say, 'Well, we've got plenty of electricity but you can't because of the environmental externalities in your countries.'" McIntire connects the lack of electricity with diminished productivity, but some experts go further, putting a scarcity of power at the root of poverty in Africa itself. The American economist Paul Romer believes that Africans do not lack electricity because they are poor: "Indeed, reliable power is so important for education, productivity, and job creation that it would be more accurate to say that many in Africa are poor because they don't have electricity."

Tanzania's parastatal power company, Tanesco, was infamously mismanaged and inefficient. In 1990 the World Bank began formulating a development aid package known as "Tanzanian Power VI Project," an initiative intended to aid the country's power company during a wider transition to a market-oriented economy. The plan, the bank claimed, would increase Tanesco's attractiveness to private investors, revamp its infrastructure, and help end the bad practices that kept poor Tanzanians without electricity. It was a loan that the bank was happy to give: Tanzania joined the World Bank Group in 1962 and over the years received $6.2 billion in credits. The Kihansi Hydropower Project in the Udzungwas constituted a major portion of this new plan; once completed, it would increase the country's power capacity by more than 40 percent. It was Tanesco's bulldozers, paid for by the World Bank, that the biologists scrambled to work ahead of at Kihansi.

Kim Howell and the team of biologists conducting the new EIA lived in tents in the forest during their field visits to take surveys of the flora and fauna. Not far from their camp, the hydropower project grew in scale, bringing thousands of people from around the world. Chinese workers built the access road from the bottom of the mountain to the future dam site. The labor camps operated like small cities, with dispensaries and pubs full of Italian, Portuguese, South American, Spanish, and Swedish workers. The Norwegians were in charge of the man-made underground powerhouse. The Mauritians ran the canteens. The South Africans worked on top of the mountain, digging the 1,300-foot vertical shaft that would reroute the water into the heart of the mountain toward turbines that would convert the energy of the drop into electricity. They chipped away at the rock with hand tools and sent the debris up in buckets. The construction of the dam had an immediate environmental impact: thousands of Tanzanians flooded to the area as though it were a gold rush, hoping to benefit economically from the project but pushing out wildlife at the same time. "The floodplains used to be filled with hippos and now there are none," said Steinar Evenson, a Norwegian engineer who worked on the dam. As he spoke he mimicked someone holding a gun and firing round after round. "They killed them all but nobody cares. They are too big. And too dangerous." Game wardens brought in from Mkumi National Park, he said, killed three male lions during construction in order to protect the growing villages of people.

Meanwhile, the team of biologists set bucket traps in the ground to catch snakes, mice, and amphibians, but they found no new or endemic species. Everything that fell in their buckets could be found somewhere else in the Eastern Arcs. The ever-present backdrop to their toiling was the one place they couldn't get to: the waterfall itself. This inaccessibility was a frustrating reminder that whatever new assessment they issued was going to be incomplete. "From an ecological point of view, the base of the waterfall has to be the most significant part of the entire gorge," said Peter Hawkes, a South African entomologist who worked on the EIA. "We went there at the end of the dry season and you could still see this massive cloud of spray." In hindsight, the biologists have conflicting theories about why they couldn't reach the spray zone. Hawkes said their local guides didn't want the biologists to see illegal logging by villagers that was taking place, and intentionally misled them. Howell said he doesn't think there was ever a clear path to get there. "You could hear it and you could see it," said Howell of the waterfall. "I have in my notes that twice I fell trying to get into the spray zone."

In December 1995, the team issued their assessment, a three-volume tome. "We didn't feel too bad about the dam. Yes, we're going to lose some stuff if the gorge dried out," summarized Howell. "But our caveat was that we could not get into the spray zone." Despite the completion of the EIA, Gerstle believed that the team should continue to meet at Kihansi in order to conduct longer-term monitoring. In December 1996, they assembled for a planning workshop and settled into camp. Gerstle suggested they see how far they could get up to the waterfall, and they were amazed to discover a new path, most likely made by Tanesco in order to install a rain gauge near the waterfall.

The spray zone was unlike anything they had conceived of in their minds. "When we actually got there, the spray was so much that it basically drowns any tree within one hundred yards of it," said Hawkes. "So you have this sloping grassland meadow, sunshine, it's out in the open. It was completely different from what we'd expected, even more unique than we'd imagined." Within minutes of arriving, Howell stuck his hand into the pile of wet vegetation and pulled out a frog. "I said, 'Yes, it's a yellow frog. It has to be new.' I mean, we looked at it and looked at it, I took it back and looked

at it under a microscope I had. I felt it had to be a species new to science because I knew all the other ones in Tanzania."

The consequences of the looming commission of the dam on the frog
were immediately clear to the team. "Oh, we knew right away. We knew
right away," said Howell. "It would become extinct."

✳ ✳ ✳

What is a species worth? There have been some 900 total extinctions or extinctions in the wild in the last 500 years, according to the International
Union for Conservation of Nature (IUCN). Why do they matter? These are
the kinds of questions that form the core of environmental ethics, a philosophical discipline that sprung up in universities around Europe, Australia,
and America in the early 1970s, just as legislation regulating the protection
of species and environmental movements was growing in tandem with pro-
gressive social causes such as civil rights and women's liberation. In 1973,
the Endangered Species Act was passed in the United States, a law that acknowledged the threat to plants and animals and their "esthetic, ecological,
educational, recreational, and scientific value to our Nation and its people."

As awareness of ecological crises grew, environmental ethicists sought
to address the claim of American conservationist Aldo Leopold that the
roots of modern environmental crises were philosophical. Unfortunately,
the logic of the emerging environmental movement and the legislation for
species protection was badly flawed from a philosophical point of view. The
argument for preserving nature was its value to humans, but what if a species had no clear value? And what if a species' interests directly conflicted
with the interests of humans? There was no coherent, rational argument on
how to answer those questions. Meanwhile, almost as soon as the ink had
dried on the Endangered Species Act, the law was challenged in the Supreme
Court. A three-inch freshwater fish called a snail darter had been discovered
in the Little Tennessee River, and the fish's migratory route happened to
be in the way of a $119 million dam project, impeding the production of
electricity for people and businesses. The Supreme Court ended the dam

project, but Congress later exempted it and the snail darter was relocated to a different river. It was obvious that in the face of strong economic interests, the protection of rare, obscure species was still seen as frivolous.

The same year the Endangered Species Act passed, Richard Sylvan published a paper in Australia called "Is There a Need for a New, an Environmental, Ethic?" A philosopher and environmentalist, Sylvan believed the dominant Western paradigm of thought was characterized by human chauvinism. Throughout history, the ultimate test of perfection in nature, he believed, had been its usefulness for human purposes. Even the transcendentalists, who revered nature as worshipers would revere their cathedral, assumed this anthropocentric perspective. Ralph Waldo Emerson once wrote that nature is "made to serve. It receives the dominion of man as meekly as the ass on which the Savior rode."

In his essay, Sylvan presented a thought experiment he called the "last man argument." Imagine you are the only person on earth. The world system has collapsed. Before you die, you make sure that every living thing in the world will be destroyed and nothing is left. Would you be acting immorally? If you believe the value of nature is its usefulness to humans, the answer is no. But of course, we intuitively feel destroying the world would be horribly wrong. That feeling was the source of Sylvan's argument that an entirely new ethic—neither a primitive, mystical, or aesthetic one, nor an economic or even scientific one—needed to be articulated. "Human interests and preferences are far too parochial to provide a satisfactory basis for deciding on what is environmentally desirable," wrote Sylvan. Nature needed to have its own value that merited our moral concern, and species themselves would need to become moral objects. Later, this concept would get a name: "intrinsic value." Sylvan's paper was one of the first to articulate it, and the idea would shape the field of environmental ethics for decades to come.

The notion that nature wasn't designed for *us*, that its value might not be described in economic, scientific, or even spiritual terms, is still fairly radical. There are traces of it in John Muir, who wrote that "the world, we are told, was made especially for man—a presumption not supported by all the facts." But until the 1970s, the chauvinism described by Sylvan had rarely or effectively been challenged. Creating a defensible concept of intrinsic value became the "theoretical quest" of most environmental philosophers, wrote J. Baird Callicott.

Intrinsic value found its greatest champion in a young Presbyterian minister from the American South, Holmes Rolston III, who would become the father of the field of environmental ethics. His ideas are so monolithic today that even those who vigorously disagree with him can't avoid tangling with what has been deemed a revolutionary moral theory. For Rolston, the value of nature is objective and is independent of human values; it preexists us and will outlast us. In his 1993 essay "Value in Nature and the Nature of Value," he wrote,

> Perhaps there can be no science without a scientist, no religion without a believer, no tickle without somebody tickled. But there can be law without a lawgiver, history without a historian; there is biology without biologists, physics without physicists, creativity without creators, story without storytellers, achievement without achievers—and value without valuers. . . . From this more objective viewpoint, there is something subjective, something philosophically hazardous in a time of ecological crisis, about living in a reference frame where one species takes itself as absolute and values everything else in nature relative to its potential to produce value for itself.

Now in his early eighties, Rolston continues to write and travel around the world. In 2003, he received the Templeton Prize, an honor bestowed most recently on the fourteenth Dalai Lama and Desmond Tutu. He wears his abiding love for the natural world on his sleeve. A third-generation Shenandoah Valley preacher who grew up in Depression-era rural Virginia, Rolston studied physics, astronomy, and mathematics as an undergraduate before getting a doctorate in theology at the University of Edinburgh in Scotland. He started his teaching career in Colorado, relocating his family to become a professor of philosophy and religion at Colorado State University in Fort Collins where he taught for fifty years. He describes environmental ethics in the syllabus for his foundational course on the subject as "an adventure in what it means to live as a responsible human being in the community of life on Earth."

I wanted to talk to environmental ethicists in order to understand cases of species extinction and conservation such as that of the Kihansi spray toad. The scientist and writer Stephen Jay Gould once said, "Our efforts at conservation have little moral value if we preserve creatures only as human ornaments; I shall be impressed when we show solicitude for warty toads and slithering worms." Gould wrote that by appreciating species for what they are, and learning from them about nature's diversity, "we shall finally understand, and to our greatest benefit in both practical and spiritual terms, what Huxley called, in the language of his day, 'man's place in nature.'" But I felt my own sympathy for the cause of conservation challenged by the little warty toad in electricity-starved Tanzania. The ethicists I spoke to said I should get Rolston's view.

The first time I talked to Rolston, he had just returned home from India where he had seen Bengal tigers in the wild, an endangered species with less than 2,500 left in the world. The experience, he said, sent chills up and down his spine. "You're talking with someone who likes to see animals wild," said Rolston in his deep Southern accent. "I go down to the Denver Zoo, I kind of pity the thing. Maybe it's got habitat enrichment, but it can't roam around or hunt. A tiger in a zoo isn't really a tiger anymore because it's not doing its thing." This was a favorite refrain of Rolston's, I soon learned. It sounded simple, almost glib—animals "doing their thing." But it encapsulates a profound moral attitude. To Rolston, every living thing has a *telos*, a function in its ecosystem, the role that organism occupies in relationship to the world. Rolston's conservation ethic is based on preserving that telos, whether it's a tiger hunting in the wild or a pasqueflower blooming in the Rockies. "You have to have a care or respect for what species are, in and of themselves," he said. "They have a good of their own that is worth saving or conserving. Telos or intrinsic value is necessary to having a functioning environmental ethic."

According to Rolston, value is located at the level of individual organisms but also species, ecosystems, and evolutionary processes. For instance, the intrinsic value of a bee is the propagation of an identity, "bee-bee-bee," over time, a historical lineage carried through individuals. "The species line is the *vital* living system, the whole, of which individual organisms are essential parts," Rolston wrote in 1993. "The species defends a particular form of life, pursuing a pathway through the world, resisting death (extinction), by regeneration, maintaining a normative identity over time. It is as logical to say that the individual is the species' way of propagating itself as to say that the embryo or egg is the individual's way of propagating itself. The value resides in the dynamic form; the individual inherits this, exemplifies it, and passes it on."

By the early 1980s, intrinsic value was a part of the environmental movement's vocabulary. In 1982 the United Nations' World Charter for Nature stated that every "form of life is unique, warranting respect regardless of its worth to man." In 1992, Rolston attended the United Nations Conference on Environment and Development in Rio de Janeiro, and when the Convention of Biological Diversity was later signed, the participating states acknowledged the "intrinsic value of biological diversity." In general, the 1990s was a decade in which public fears of large-scale ecological collapse brought on by global warming, pollution, and capitalism grew, as did a concern for the environment and support for conservation causes. And though the average citizen in Europe or America wouldn't recognize Holmes Rolston III walking down the street, his idea that *all* ecosystems and biodiversity (a term coined in the 1980s) warranted protection became more mainstream. When Dave Foreman, the environmentalist and self-described eco-warrior, identified four things that were shaping the conservation movement of the 1990s, he named academic philosophy as the first, ahead of conservation biology, local environmental organizations, and his own radical group, Earth First!.

Of course, like all theories, intrinsic value has dissenters. Moral pluralists challenge the notion of a single, universal environmental ethic. Deep ecologists resist a concept of nature or wilderness as something separate from humans. Social constructivists believe that ideas about nature are culturally relative. Ecofeminists reject patriarchy that has oppressed both nature and women alike and persists in the philosophical disciplines. And there are those who see the field of environmental ethics as inhabiting an ivory tower disconnected from the actual grunt work of trying to solve real ecological crises. In the early 1990s a number of philosophers formulated a new camp and called it environmental pragmatism. Their work focused on making philosophical contributions to the resolution of "real world" crises. (One of them, Bryan Norton, feared environmental philosophers would "continue to dance with Cartesian ghosts even while tropical forests, both ancient and modern, burn.")

Despite the efforts of the environmental pragmatists, it's fair to say that the bridge between environmental ethics and the applied sciences has never been successfully completed. Today, when the pressure to *do something* about threats to species before it's too late is strong, philosophy can seem irrelevant and distracting to the task of policy making, and a downright impediment to navigating real political and economic quandaries. As John Lemons, an environmental scientist at the University of New England, wrote in an essay published in 2007: "From an environmental professional's perspective, there is neither an environmental ethic that is fully developed nor one that has been used extensively to inform policy making and decision making." Few park rangers or wildlife biologists—let alone politicians—choose to utilize philosophical concepts in order to answer the questions they face about how and when to protect species.

In the four decades since the field of environmental ethics began, the focus of inquiry has also shifted. Ecological crises are raising questions beyond those that concerned early ethicists and conservationists—such as how to preserve wilderness and landscapes—to economic disparities between the rich and poor, international human rights, and the notion of environmental justice. The fact that climate change is anthropogenic, caused by a population of 7 billion humans, and impacts even the furthest reaches of the planet, is also a game changer for the field. "The question today is do we need to revise our ethics in light of the fact that it's inevitably going to be a human-dominated world?" Rolston told me. "We have a lot of people saying we're in the Anthropocene Age, and environmental ethics has to change its focus a lot, to humans managing everything. That you can't really think of any place in one hundred or fifty years from now that's not going to be modified by global warming and there won't be such things as wild places anymore. Everything is going to be a big zoo." Unsurprisingly, he is less than enthusiastic about this prospect. Some might see managing the planet as a kind of utopic destiny for our technologically brilliant species. Not Rolston. "We haven't been good at managing bits and pieces of it," he said. "I want wild nature today as near as we can have it. Animals doing their thing. Natural selection continuing to take place."

As crises unfold with more species coming closer to the perilous edge of demographic collapse, many of the moral puzzles intrinsic to our modern relationship to nature are going unsolved. Should the conservation of a species ever be put above the needs of humans? How far should scientists go to prevent the extinction of a species? Can a species ever be wild again after we save it?

These are just some of the dilemmas that were set in motion when Kim Howell pulled a small yellow frog from the base of a waterfall in the Tanzanian rainforest.

* * *

Nectophrynoides asperginis, as it would be called two years later when it was formally described in the *African Journal of Herpetology*, is not particularly colorful, or cute, or weird looking. It has matte, mustard-colored skin and is very, very small, with a body about the size of a nickel. Rather than depositing eggs that then grow through a tadpole stage, spray toad mothers give birth to fully formed babies. The purplish-hued toadlets are so little they can fit on the tip of a pen. By far the most fascinating thing about the spray toads was their adaptation to their habitat: five square acres of wetlands, where the combination of spray and wind created by the power of falling water hitting rocks produced a microclimate no one had ever seen before. Later studies of the volume generated by the falls found that around three-quarters of a million liters of spray was created each day. In order to hear one another within the roar of the falls, the toads developed a unique form of acoustic communication. Spray toads have no outer tympanum; in other words they are earless. But they do have an inner ear that is capable of detecting ultrasonic noise—a frequency above the upper limit of human hearing. Biologists believe the frog's combination of ultrasonic and visual communication was a survival mechanism that helped them find reproductive partners, despite living in an environment that was the equivalent to a deafening rock concert for humans. The toads' perfect adaption to the wetlands also meant they were totally dependent on the millions of fine droplets of water that created a constant shower in the middle of the forest. "Most toads are tough," said Howell. "That's the classic amphibian. Amphi, two. Bios, life. Two ways of living: outside the water and inside the water. But Kihansi spray toads aren't like that. Unfortunately for it."

Subsequent visits to study the new species showed there were thousands of toads. Protected by the intense spray, the species had no serious predators, not even birds from the surrounding forest. But the spray meant conditions for the researchers were extremely difficult. "It was scary," said Charles Msuya, Tanzania's foremost herpetologist, who has studied in the Eastern Arcs for decades. Navigating the steep, wet rocks was treacherous. The ground underfoot was soft and spongy, almost gelatinous; stomping on it would cause the ground to vibrate a yard away. Furthermore, because of the cold temperature of the water coming over the falls, field visits were freezing. "I used to walk into the shade of the forest to warm up," said Peter Hawkes. "We tried everything, wet suits, semi-wet suits, six layers of jerseys, waterproofs on top of wet suits, anything you could imagine, but spending a day in the wetlands you were cold. And outside it would be ninety-five degrees."

According to John Gerstle, "As soon as we identified [the toad] we notified the World Bank and Tanesco and engineers that were responsible and said, 'We can't move ahead without some conditions that ensure the survival of the species.'" But everyone knew that rare endemic toad or not, there was no stopping the dam, scheduled for completion in 2000. Construction carried on to the drumbeat of Tanesco, who at the time was load shedding—shutting down electric power so the system wouldn't become overwhelmed by demand—because of a countrywide drought. As early as 1998, Norway's development agency Norplan recognized the toads could become a significant problem and began investigating ways in which they might be saved. Executives at the agency discussed options such as translocating a portion of the population to another site, installing an artificial spray system, captive breeding, or allowing a sufficient bypass flow over the waterfall that would sustain the frogs. Tanesco only had a provisional water right to the falls, granting them permission to divert the river for the dam while maintaining a bypass flow of 7.7 cubic meters per second. But the company contested the amount and had been lobbying for a much smaller minimum bypass of 1.5 cubic meters per second. Every drop of water diverted for the frogs was electricity and money lost. When Norplan went to Tanesco with suggestions, the power company said that the uniqueness of Kihansi's ecosystem was grossly exaggerated.

Ironically, just a few months after the discovery of the new species, Tanzania had ratified the Convention of Biological Diversity, which legally committed signatories to protect biodiversity within their borders. This kind of legislation and the act of environmental protection itself is controversial in East Africa, a region where the history of conservation is rife with foreign paternalism, racism, and colonial humiliations. In 1903, British colonial officers in East Africa started one of the first organizations in the world dedicated to protecting animals. Called the Society for the Preservation of the Wild Fauna of the Empire (now the Fauna and Flora Preservation Society), its purpose was to protect animals in designated parks for recreational hunters, which effectively kept subsistence hunters out, according to anthropologist Janet Chernela. This eventually led to the creation of Tanzania's Serengeti National Park in the early 1920s (the Masai were evicted from the park in 1959) and, as Chernela explains, established the precedent and groundwork for agencies such as the IUCN, and the Convention on the International Trade in Endangered Species of Wild Flora and Fauna (CITES). Thirty-two percent of Tanzania's land is protected as parks and reserves, among the highest percentages in the world. But the country still struggles to strike a balance between land protection and land use, and *who* gets to use it. For instance, in 2014, the Tanzanian government announced plans to establish a thousand-square-mile tract of land east of the Serengeti as a "wildlife corridor" that would be off-limits to Masai pastoralists who use the land for cattle grazing, yet available to a United Arab Emirates–based safari company.

For these reasons and others, there is a strong suspicion and cynicism when it comes to the protection of nature in East Africa. Whose interests does conservation serve? In the last few years, some conservationists have attempted to address the notions that biodiversity conservation is out of touch with the realities of poor communities and that national parks and protected wilderness only serve the interests of a global elite. They seek to promote poverty alleviation and economic development as a conservation tool, at times by advocating forestry and agriculture in the belief that a higher standard of living for the poor will ultimately benefit nature conservation. Sometimes called the "new conservation," it has came under fire by the old guard such as Michael Soulé, one of the founders of the field. As he wrote in the journal *Conservation Biology* in 2013: "There is no evidence for the proposition that people are kinder to nature when they are more affluent, if only because their ecological footprints increase roughly in proportion to their consumption . . . I must conclude that the new conservation, if implemented, would hasten ecological collapse globally, eradicating thousands of kinds of plants and animals and causing inestimable harm to humankind in the long run." Soulé's fear is that humanitarian-driven conservationists will only demand the protection of nature if that nature is materially valuable to people. His concerns seem more than justified, and yet there is a terrible tension in the fact that so often it is white Western scientists demanding that developing countries like Tanzania not do as the scientists' own countries did for generations.

In the case of the rare Kihansi spray toads, their protection seemed to be in no one's interest, at least initially. The World Bank wasn't willing to act on the emergency mitigation measures recommended by Norplan and refused to formally acknowledge the discovery of the toad until a donor meeting held in March 1998, where Howell says he was treated as an alarmist. "They said to me directly, 'How do you know this isn't found somewhere else?' . . . We had immediately looked in other places of course. That's the first thing you do as a scientist. You want to know, is it really only found in one little area the size of a football pitch? Or is it found in other areas? So we looked and we couldn't find it. And the reason is clear. It's the same reason that Tanesco chose the Kihansi River, because it has constant flow even in the dry season." For Howell, the actions of the World Bank were unpardonable. He felt their lack of concern reflected a flippant attitude toward following inconvenient rules in an African country.

Construction of the hydropower project continued. Then the bank's attitude suddenly changed. In November 1999 a letter from the environmental group Friends of the Earth arrived for then-president James Wolfensohn. Someone had leaked the group information about the frogs and they pointed out that the World Bank was in danger of violating its own environmental policies. It was the beginning of a potential public relations nightmare, and the conservation of the toad quickly became very important to the bank.

It was into this fray that Bill Newmark, a prominent American biologist, arrived in 2000 as a consultant for the World Bank. Newmark is an expert on wildlife corridors, ways of connecting protected areas like national parks in order to facilitate movement between conservation "islands." He first gained widespread respect as a graduate student in the 1980s, studying in the western United States, where he found that wildlife reserves were not protecting biodiversity as predicted but actually losing species, particularly mammals, because the parks were too small to support them. His findings appeared in *Nature* in 1987, and Newmark went on to become an influential conservationist. The World Bank hired Newmark—who had worked extensively in Africa to create corridors between national parks for elephant migration and studied bird extinction patterns in Tanzania—to figure out how to save the spray toads. Newmark told me that when he first began coming to the Eastern Arcs over twenty-five years ago, "it was the most remote and poorest area I'd ever been in. It had not been visited, it had not been surveyed. That's why we're finding all these new species."

In Newmark's opinion, "you could not have designed a more ecologi-cally friendly dam. All the water is returned to the river, it's only diverted out of a five-kilometer section of the river. But it turns out, that five-kilometer stretch was where the spray toad was located." Six weeks after all three tur-bines of the dam turned on in 2000, Newmark found that the spray zone at the waterfall shrank 98 percent. Inside the gorge, he saw toads huddled in clumps near the bases of the falls. Within weeks the estimated toad popu-lation dropped from 20,000 to 12,000. Newmark, whose findings were re-ported directly to the office of the World Bank president, recommended immediately installing the artificial sprinkler system and creating a cap-tive breeding population. Work began on the sprinklers—an ingeniously simple gravity-fed irrigation system of rubber hoses and dozens of sprinkler heads—but the Tanzanian government refused to issue the export permit needed to begin a captive breeding program. The fear, according to Anna Maembe, an environmental official in the government, was that Tanzania would lose control of a potentially valuable natural resource if the frogs be-came the basis of a vaccine or drug. "We were of the opinion, by 'we' I mean Tanzania, that we would rather have it here and do our studies here, rather than someone keep it for us," she explained.

In the United States, the Bronx Zoo (owned by the Wildlife Conser-vation Society) and the Toledo Zoo offered to maintain toad populations. But Tanzania held firm until, according to one insider, threats were made that if Tanzania did not cooperate, future World Bank funding for devel-opment projects could be at risk. 2000 was an election year and Tanzania's then-president Benjamin Mkapa received a call from a bank official saying, "I would hate for the president of the World Bank to call you and instead of congratulating you on winning the election, ask you about these toads." The government granted an export permit and Mkapa won reelection. Af-terward, Jason Searle, a Bronx Zoo biologist, arrived in Tanzania to airlift the toads out of the gorge. "A lot of Tanzanian politicians wanted to know: 'What's the big deal? You are weighing these tiny little toads against power to our people,'" he told me. "I don't think anyone is going to argue that these toads are more important than providing electricity." At Kihansi, Searle collected 500 toads and flew them back to the United States in boxes lined with tinfoil and wet paper towels. Only one toad died along the way. The Tanzanian public did not welcome the start of the captive breeding program, funded by a loan through the World Bank. A newspaper article asked, "Is it worth paying all that money for some tiny reptiles called spray toads when thousands of Tanzanian under-five kids, pregnant mothers and retired senior citizens are dying of want?" Even some conservationists grumbled: "I think many groups would have liked to see this toad go extinct, because they finally had a real issue for legal action against the World Bank," one told me.

✳   ✳   ✳

As a conservation measure, captive breeding is controversial, though nothing new: zoos and aquariums have served as custodians for endangered species for decades. The California condor, whooping crane, black-footed ferret, Arabian oryx, and Chinese crocodile have all avoided imminent extinction because of biologists who brought them into controlled settings, where they could manage their populations and genetic stock. After the 1973 Endangered Species Act, "a lot of people became very interested in doing more than just managing collections as zoo curators," said Chris Wemmer, a former scientist emeritus at the Smithsonian's National Zoological Park. "There was a groundswell of enthusiasm for the idea that we could save wildlife." However, in many cases, said Wemmer, advocates for captive breeding became overkeen, and "there wasn't careful assessing of the needs of these different species; the goal was just captive breeding. It really pissed off people who were concerned with natural habitats and the animals in the wild."

The "arks" themselves aren't always effective. The loss of genetic fitness—measured as the number of offspring produced by individuals that survive to reproductive age—can be rapid in captive populations, occurring within several generations and leading to lower reproductive rates and higher offspring losses. Captivity can create selections of traits within populations that improve their survival in captivity but not necessarily in the wild—if they are ever returned at all. Although most captive breeding programs are intended to produce animals for reintroduction to their habitat, there are few successful cases to point to in which captive animals have become truly self-sustaining or "wild" again. Whooping cranes, for example, must still be taught to migrate by human pilots. When it comes to amphibians, the success rate for reintroductions is particularly low. One study showed that of fifty-eight species that were reintroduced after captive breeding, only eighteen bred successfully in the wild, of which thirteen were self-sustaining. Most tellingly, out of 110 species in captive breeding programs, 52 of the programs had no plan in place for reintroduction at all—their ecosystems didn't exist anymore. Proponents of what's known as *in situ* conservation, preserving animals in their native place, say this is the most destructive aspect of captive breeding; by trying to minimize the chance of extinction, it emphasizes saving animals rather than environments. "We'll never know with any degree of certainty whether these animals can be reintroduced or not," said Mark Michael, a professor of environmental ethics at Austin Peay State University in Tennessee. "There are a lot of environmentalists who say, 'If you take a species out of the wild and there is very little possibility of reintroducing them, then you shouldn't do it.'" But proponents of captive breeding believe it's better to have the species in the world than to let them disappear, even if the animals that remain in zoos are essentially, as Michael described, "museum pieces."

The spray toad captive breeding program began like many others—as an insurance policy in case something went wrong with the wild population. Initially, the toads in the gorge rebounded after the artificial spray system was installed—going from around 1,250 to over 17,000—though it required daily maintenance by a team of at least half a dozen monitors, local Tanzanians that brought their supplies into the forest and set up a permanent camp. There were so many toads, Newmark recalled, that surveyors had trouble not stepping on them. Back in the United States, the zookeepers were under pressure to maintain the captive population as the zoo toads began to die. "Amphibian medicine is so far behind bird and mammal medicine," said Jenny Pramuk, a herpetologist who ran the Reptile House at the Bronx Zoo. The toads sickened easily, and zookeepers realized that the normally fecund toads weren't procreating. They were kept away from other species in biosecure rooms that zookeepers entered only after sanitizing themselves. Months were spent troubleshooting, changing water filtration systems, food, and finally the light bulbs. Ultraviolet bulbs boosted the toads' vitamin levels and they started reproducing. But the overall captive population had plummeted to as few as seventy individuals, shrinking the species' gene pool. Small populations of animals are at the mercy of chance. Their ability to survive and adapt to disease and adverse conditions is compromised. They can begin to suffer from what biologists call inbreeding depression, a reduction in reproduction and survival, and carry a higher genetic load, the amount of harmful genetic material within the population.

As the zookeepers struggled to keep the captive population alive, the worst-case scenario unfolded at Kihansi. "We always wondered what would happen if the spray irrigation failed," said Newmark. What was the minimum amount of water that needed to be diverted in order to maintain the habitat? In June 2003, two tests were conducted, allowing gushes of water to rush over the falls again. A week later the spray toads' numbers began to decline. By July, their number had dropped to about 150. In August, only two could be found. Everyone's fears had been realized; the toads had gone extinct in the wild. What happened exactly remains a mystery. One hypothesis is that flushing the water released sediment contaminated with pesticides from upstream farms. Someone found safari ants in the spray wetland and speculated that they had preyed upon the frogs. But the most likely candidate was the most perplexing—an outbreak of a mysterious disease called *Batrachochytrium dendrobatidis,* otherwise known as chytrid fungus or "Bd." Amphibian experts became aware of Bd in the 1990s, but it wasn't formally described until 1999. About a thousand variations of the fungus exist in moist soil and leaf litter around the world; only one affects amphibians. Bd attacks amphibian skin, causing it to harden and thicken. Amphibians transmit oxygen, sodium, and potassium ions through their porous skin, and when their skin clogs, their hearts stop.

The Global Amphibian Assessment lists over 400 species as critically endangered and 122 as possibly extinct, many because of the fungus.

"We're talking about whole members of an assemblage vanishing," said Brian Gratwicke, a research scientist at the National Zoo in Washington, DC. "We're in the same place as everyone else: elephants, pandas, the fluffy stuff. Some people want to put all frogs in the same basket, but there are 6,000 species of frogs. If you want to compare apples to apples, you can put a Panamanian golden frog in the same basket as a panda." The only countries where Bd has been tested for and not been found are New Guinea and Borneo. (Until 2015, scientists hadn't found it in Madagascar.) On epidemiological maps of Central America, the spread of Bd looks like a tsunami: creeping north and annihilating species, sometimes within weeks. Some scientists have linked global warming to the sudden deadliness of Bd by linking changes in air and sea surface temperatures to the timing of population losses. Maybe warmer temperatures compromise the frogs' immunity to the fungus? Maybe they make the fungus more lethal? Some have linked Bd to drought. When it comes to this disease, no one is sure. "It's just rewritten the history books on disease epidemiology," says Ché Weldon, a South African herpetologist.

Weldon believes that pinpointing its origins can help solve the mystery of Bd. To do this, he began testing amphibian museum specimens in South Africa and found that even a hundred years ago, it was present. "Not only that, there's very little, almost no evidence of a species in South Africa becoming diseased. So if the animals have got resistance, what does that tell you? It tells you that the animals have co-evolved with this disease." Weldon thinks the most likely candidate for spreading the disease from Africa was *Xenopus laevis,* an aquatic, claw-toed frog that was used for pregnancy tests in women; when injected with urine from pregnant women, the frog would begin ovulating. When this was discovered in 1934, the frogs were caught by the tens of thousands in the wild and shipped around the world. Over a period of forty years, feral populations of *X. laevis* were established in Britain, the United States, and Chile. "It's part of a jigsaw puzzle that's not yet complete," Weldon told me. "A lot of pathogens are *species* specific. And there's absolutely no, no specificity in amphibians regarding chytrid fungus. In that sense, it's absolutely unique."

In 2007, a herpetologist named Kevin Zippel responded to the epidemic by launching the Amphibian Ark. Zippel wants to quarantine 500 endangered frog species until a solution to Bd is found. He is attempting to raise at least $50 million—an incredible sum for amphibian conservation. This quarantine strategy is changing the conversation around captive breeding. For Zippel and many other conservationists, there just isn't enough time to answer ethical questions about *ex situ* versus *in situ* conservation or whether a species merits dramatic intervention. "We're the first ones to admit this is the last-ditch solution," Zippel said.

※　　※　　※

On a Saturday in spring, I rode in a car with the Tanzanian herpetologist Charles Msuya from Dar es Salaam to Kihansi. Msuya explained how biologists didn't explore the area known as Sanje, a region of the Udzungwa Mountains, until the 1980s, like much of the mountain range. (The name Eastern Arc Mountains wasn't used until 1985.) They quickly discovered a new endangered species of monkey, the Sanje mangabey, and subsequently the government of Tanzania was pressured to preserve the forest by creating a national park. The move shut out the villagers who used the forests for firewood, medicinal plants, rituals, and burials. Today, the government has attempted a compromise, opening up the forest one or two days a week for people to take what they need. Some villages also have access to a plot of forest where they can maintain a family burial place.

After ten hours, seven of which were spent on a pocked dirt road the color of ground cinnamon and seemingly as soft, a sharp right turn directly into the foothills brought us to a paved road. Ascending higher into the mountains, well-built cement homes appeared, and finally a sprawling guesthouse with swimming pool, bar, and flat-screen TV showing English soccer. The facility houses Tanesco employees and their families, engineers, and foreign visitors. In a couple of days, biologists, including Bill Newmark and Jenny Pramuk, would be flying in to a small airstrip nearby and we would all hike up to the gorge together. After dinner that night, I found myself talking to a Norwegian engineer, Steinar Evenson, in his early fifties. "Have you come to look for little frogs?" he asked. "That is an expensive frog. The most expensive frog in the world." I told him I wasn't here to look for frogs myself, per se, but rather to look at the people looking for frogs. It turned out that he had been working in East Africa for three decades and lived at Kihansi for three and a half years in the 1990s, when the dam was being built. "What is so special about this frog?" he said. Nothing much, I confessed, except for the waterfall. Evenson seemed disgusted. "Who is paying for all these scientists and biologists to fly here and look for this little frog? It's crazy. They should have used all this money to build a fourth turbine."

The next day I rode along in an SUV to the nearest town, called Mlimba, where supplies were bought for the biologists' arrival. At some point I lost the translator, so I spent the next couple of hours sitting on a narrow wooden bench under a tree with the driver and a young boy, with whom we shared the roasted ears of corn and pineapple wedges we bought. United in our boredom and forced silence, we watched the village life around us: a group of men playing checkers in the shade; a woman putting braids into a young girl's hair; an elderly dwarf also eating corn. A man who looked to be in his nineties with yellowed hair trudged past us slower than a tortoise. Across the street I spotted a young mother holding a newborn baby to her chest. When we eventually left, she climbed in the SUV for a free ride. She had come the twelve miles from her home to buy cough medicine. Her purse, a plastic potato sack with two holes cut into the top for handles, contained blankets and, meticulously wrapped in newspaper on top, the box containing the medicine.

As we drove back toward Kihansi I caught my first glimpse of the waterfalls. At that moment, two of the three turbines at the power plant were shut down for repairs, so the volume of water flowing over the falls was relatively large, much closer to what it would have looked like before the dam was built. I had imagined them to be hidden, a secret trickle tucked away in the forest, but there the water was shining in the sun, visible from miles away. The falls looked powerful and huge.

The next morning the team of biologists arrived, and after lunch we started the trek, decontaminating our hiking boots in shallow buckets of bleach at the foot of the trail, an attempt to prevent any pathogens from entering Kihansi. It was still unknown how Bd had gotten into the gorge in the first place. In 2007, at a particularly sensitive meeting on the Kihansi conservation plan held in Tanzania, a Tanzanian biologist had suggested that American herpetologists had brought Bd with them into the gorge. The entire world population of the spray toads was around 500 (today it is around 6,000 in captivity) and being maintained in the United States; the Americans were furious at being implicated in their extinction from the wild.

The Udzungwas are home to an incredible amount of biodiversity: butterflies, centipedes, snails, bees, ants, hornbills, and primates. As we hiked up the steep trails, porcupine quills and crab shells discarded by African clawless otters littered the ground. We crawled over fallen tree trunks and around neon-green, moss-covered boulders, vestiges of the Precambrian eon over 3 billion years ago, or as Newmark described it, "some of the oldest rock in the world." A few hours later, we arrived at the Kihansi research station, a spacious, Adirondack-looking cabin painted green with wrap-around porches, where a group of Tanzanians live and collect data from the gorge, and maintain the artificial spray system, which runs twenty-four hours a day.

Newmark told me the gorge is "probably the most highly engineered recovery plan for any species in the world." The question on everyone's minds was this: Would the habitat, after nearly a decade of artificial maintenance, be capable of maintaining a population of toads in the future? A reintroduction would be an unprecedented victory for amphibian conservation, but the costs could run into tens of millions of dollars. The captive frogs might carry an unknown pathogen into the gorge, or they might die off outside controlled conditions. A decade of captivity has undoubtedly led to unintended natural selection for characteristics beneficial to surviving in their zoo environment. "The Kihansi spray toad was down to seventy-two individuals, so that's the genetic snapshot," said Pramuk. "From there, it's artificial selection really because we selected for our tanks, our water. I think our population [at the Bronx Zoo] even looks a little different from Andy's [Andy Odum, the spray toad keeper at the Toledo Zoo]. We know we're not going to restore the exact same genetic pool to the gorge, but we are restoring the ecology of the gorge. Maybe they're better adapted, maybe not, maybe they're less. We won't know till the moment of truth."

That night Pramuk, along with Tim Herman and Andy Odum, the Kihansi spray toad keepers at the Toledo Zoo; Chris Hanley, a veterinarian; David Miller from the US Geological Survey; Mutuguaba, a Tanzanian gorge surveyor; and I put on yellow rain slickers, rubber boots, and headlamps and headed out into the forest. It was a slow-going hike on the narrow path, and everyone was stopping to shine their flashlights into the foliage and debating the taxonomy of what they saw. Odum wanted to find a deadly green mamba or a python, but among the many geckos, chameleons, stick bugs, grasshoppers, snails, and spiders that we saw, the only snake he spotted was a white-bellied specimen, roughly two feet long, with a green stripe on its head. Pramuk identified it as a rear-fanged type but no one was sure what species. When we found a second one hanging in a tree, I leaned in about a foot away and took a picture.

Up to this point, I hadn't understood why we were dressed like the Gorton's fisherman. The closer we got to what's known as the lower spray wetland of the Kihansi falls, the long-ago habitat of the spray toad, the more obvious it became. I was drenched before the roar of the falls, which were invisible in the pitch-black and beyond the scope of our flashlights, became so loud that it drowned out our conversation. We had reached the artificial misting system with rubber hoses running up and down the soggy and steep slopes of the gorge. The ground beneath us was like pudding, so soft that our boots sank six inches into the trembling earth. Mutuguaba, who manages a staff of seven that monitors the sprinkler heads on a daily basis and records data such as humidity and temperature, stepped between two rocks and fell through the soft ground up to his waist. Judging by the noise of the falls, which were "running" at a fraction of their former force, the experience of standing next to them before the dam was built must have been terrifying.

All of us shined our flashlights onto the glistening boulders where the spray toads would have once congregated in the thousands. "What? No spray toads?" said Tim Herman with mock surprise.

Back at camp, Odum and I sat down at the kitchen table and opened up a field guide to the amphibians of East Africa, co-authored by Kim Howell. He sucked in his breath and pointed it out to me. "*Thelotornis kirtlandii*," he said. "That snake has killed people. I can't believe I didn't identify it. That's a snake of consequence." I read the description: "The venom is potent, causing a general bleeding tendency. No anti-venom is produced." "What does 'a general bleeding tendency' mean?" I asked. "It means you hemorrhage from every orifice until you've bled to death," he said.

The next morning everyone put on raincoats and rubber boots again for the climb to the gorge. The tree cover opened up and the towering rock faces of the gorge appeared, rising into the low-hanging cloud cover. As we stood in the artificial sprinklers, our feet sank and the relentless spray drenched us again. The fate of the Kihansi spray toad and the precedent for other conservation interventions rested on whether or not the frogs could be reintroduced to this place one day. Newmark seemed pleased—some of the plant and shrub species that had invaded the wetlands when they dried out were less pervasive. Jenny Pramuk appeared melancholy to me. She stared from beneath the hood of her blue raincoat at an enormous boulder where thousands of toads once evolved. "What about the temperature?" she asked no one in particular. Without the floods of cold water coming over the falls, the gorge itself had warmed. Could the toads adjust? Had the environment itself gone extinct?

"If the reintroduction doesn't work, at what point are we just going to say good-bye?" she said.

✳ ✳ ✳

It was an awful long time before the moment of truth arrived. In July 2012, three years after I hiked with Newmark and Pramuk into the falls, biologists brought the first frogs back to the Udzungwa Mountains and released them into the artificial spray zone. The media treated their return to Africa as a historic occasion, the first successful reintroduction of an amphibian species to its wild habitat. In reality, the toads' return to Tanzania was complicated and their success in the forest far from guaranteed. Biologists put the first population of frogs into cages where they could be monitored and protected from predators. Later, 2,000 were released and biologists marked a quarter of them with dye in order to track their survival rate, which appeared to trail off. Adult frogs in particular seemed to struggle. "There is the possibility that these adults have led their whole lives in Bronx and Toledo zoos for the last fifty generations; maybe we are bringing them back and these animals aren't able to adapt as well even if it's their native habitat," said Kurt Buhlmann, a biologist involved in the reintroduction effort. Captive-bred frogs will most likely supplement the population for years to come. "There's no cookbook for this," said Buhlmann. "We've always stressed that it may not work on the first time. And even if it works a little bit, it's going to require reinforcements over and over."

In the years since the field of environmental ethics was born, there have been many different arguments for why we should preserve wilderness and the species within it. Some see nature and species as natural resources, others as having potential pharmaceutical value. Nature provides important services to humans—cleaning our air, say, or providing us with places for physical exercise and mental rejuvenation. Species may be aesthetically beautiful and inspire us individually and culturally, or represent transcendent moral truths. They are records of evolution containing information we need in order to understand life on earth, or their value is their wonder for the human intellect. Wilderness is where humans come from, it's our history as a species, we need it in order to understand the relationship between self and the world. Some believe wild things have the right to exist without human disturbance, or that we should protect them for future generations. For Holmes Rolston III, the reason to protect wilderness and species is because they exist. Their preservation should need no justification, because their value is intrinsic and independent of human perspective. But sometimes, like with the spray toads, it's in the protecting itself that the trouble can begin.

I recalled a conversation with Kim Howell, now seventy years old, after I returned from Kihansi. We met on the run-down, verdant grounds of the University of Dar es Salaam, sitting outdoors at a concrete table near a frangipani tree. He still spoke with disdain about the World Bank's actions. "I've often said I wish I had never discovered the toad," he said. The species' removal for captive breeding allowed the bank to continue with the dam, ultimately destroying the integrity of the entire ecosystem. "Any development in the East Arcs is going to involve the loss of an endemic species," Howell said. "Forget about invertebrates—we know from my work on millipedes that every mountain has its own fauna. Maybe even every forest has its own fauna . . . I've been here forty years, I know there are many, many more Kihansis coming."

2

TRACKING CHIMERAS IN THE FAKAHATCHEE STRAND

Puma concolor coryi

The puma is a cat with many names. Catamount, cougar, mountain lion. In the southeastern United States, pumas are called panthers, and in the early 1970s there were two camps among biologists when it came to the question of whether any were still living. The first camp believed the subspecies had completely disappeared. Colonizers in this part of the country beginning with the Spanish conquerors saw the animals as a menace and killed them accordingly. Bounty laws awarded anyone who killed a panther and could prove it with a scalp; in 1887 a dead panther was worth $5. Meanwhile, hunters, squatters, and sprawling agriculture wore down deer populations, the staple of the panther diet. The result was that by the end of the nineteenth century *Puma concolor coryi*, the sleek animals that had once roamed from the bottom of South Carolina to Tennessee,

Arkansas, and Louisiana, had nearly vanished. "The exact range of the form cannot now be given, as the puma is extinct in all the region directly north-east of Florida, and I believe northern Florida as well," reported a member of the Boston Society of Natural History in 1898. "None have been seen in eastern Georgia for many years."

The second camp thought that some cats—maybe as many as 300—might have found a way to survive off feral hogs in the forests and swamps of southern Florida, places where agriculture was near impossible in the acidic soil and development was limited by the brutal heat and humidity of the tropical landscape. There was some evidence to support this idea. In 1969, a deputy sheriff killed a 100-pound male panther near the town of Inverness in central Florida. Three years later, a highway patrolman shot and killed a pan-ther east of Lake Okeechobee after a car injured it. Based on these encoun-ters, there was reason to believe there could be a larger remnant population of the subspecies still eking out an existence in Florida's wilderness.

Around 1972 the World Wildlife Fund (WWF) decided to find out. It was just one year before Congress would pass the Endangered Species Act. Florida law had fully protected the panther from hunting since 1958, but would the cat be listed under the new federal law as threatened? Endan-gered? Extinct? The environmental organization reached out to a taxono-mist in Florida, who in turn contacted a predator hunter and tracker in Texas by the name of Roy McBride.

McBride was an unlikely character to be hired by a group of conserva-tionists interested in the survival of panthers. The author Donald Schueler in his 1991 book, *Incident at Eagle Ranch*, presented a rare portrait of a private man who avoids the public eye. "During his younger days," wrote Schueler, "he had more to do with bringing the mountain lion to the verge of extinction in Texas than any other single person. Given his remarkable stamina and the quality of his pack of hounds, a lion 'almost never gets away' once McBride goes after it." "Let McBride do it" was the motto when a particularly wily predator was amok and needed capturing.

McBride used a variety of means to take an animal. If he didn't have an effective tool at his disposal, he invented it. In the 1970s he was having difficulty trapping a coyote on a sheep ranch. He imagined that if he could put a trap where the animal attacked the sheep's neck, he had a foolproof way of catching a hungry coyote. Of course he couldn't attach a trap so instead he created a collar with poison that could be attached to the sheep's neck and would kill the coyote. McBride turned his patented collars into a family business near his home in Alpine, Texas. When President Richard Nixon signed an executive order restricting the domestic use of the poison known as Compound 1080—a popular substance with no odor or taste used for decades to poison large predators, and used in McBride's collars—he sold them to ranchers in Mexico, Canada, Argentina, and South Africa.

In addition to his work in the United States, McBride had traveled for years throughout Mexico as a contracted wolf hunter. He spoke Spanish fluently and rode on horseback to track the animal called "El Lobo" for ranchers who needed to protect their livestock. When Cormac McCarthy was writing *The Crossing* in the 1990s, the second book of his Border Trilogy series, he found inspiration for the story of a haunted sixteen-year-old boy trying to catch a wolf in McBride's tale of spending eleven months hunting a single wolf in Mexico. Among hunters and naturalists in the Southwest, the story is the stuff of legend.

The male wolf called "Las Margaritas" had lost two toes on its left front foot from an encounter with a trap. In the late 1960s, he was killing dozens of yearling steers and heifers on ranches along the Durango-Zacatecas border. "The wolf seldom used the same trail twice and if he came into a pasture by a log road, he left by a cow trail," wrote McBride in a government report in 1980. "I was sure I could catch Las Margaritas, but I couldn't get him near a trap." McBride tried baited traps and blind traps, traps boiled in oak leaves, and traps concealed in carefully sifted dirt. Nothing worked. Over months of intensive effort, McBride had managed to get the wolf close to a trap just four times. He traveled thousands of miles on horseback trying to understand the animal's uncanny ability to elude him. "Almost a year had passed and I was now convinced that I would never catch this wolf," he wrote. "Just how the wolf could tell the traps were there is something I cannot comprehend to this date." At times, however, McBride had noticed that Margaritas had passed at campfires along the road, places log-truck drivers had stopped along the way to cook. "I set a trap near a road that the wolf was sure to come down if it continued to kill in the area, built a fire over the trap and let it burn itself out." McBride put a piece of dried skunk hide in the ashes from the fire and waited. One day in March, the wolf caught wind of the setup and went to investigate. The trap caught him by his crippled foot.

Wolf lovers and conservationists might shudder at the story of a man hunting down what we now know was one of the few remaining individuals of the Mexican gray wolf species in the wild. But McBride's legacy is more complex than that. In 1976 the Mexican gray wolf was listed as endangered under the Endangered Species Act, and the US Fish and Wildlife Service (USFWS) Office of Endangered Species hired McBride to find out whether any of the wolves the government had once tried to eradicate had survived in Mexico. He found some twelve wolves in the state of Durango and half a dozen in the state of Chihuahua. In all, he estimated as many as fifty individuals in all of Mexico might still be living, but the possibility of the species being saved in the wild was, in his opinion, impossible. The next year he trapped six gray wolves—two in Sierra del Nido in Chihuahua and four near Coneto in Durango—and delivered them for a government captive breeding program in Tucson, Arizona, with the goal of bringing them back to the landscape. "It was a change in policy, to say the least," he told me. "They were the guys who killed them and then they tried to reintroduce them." After years of political controversy, bureaucratic turmoil, and fluctuating populations, around eighty Mexican gray wolves roam the Southwest today, more than at any time since the government reintroduced them to the wild in 1998. These wolves descend from just seven wolves representing three captive lineages: the Aragón, Ghost Ranch, and McBride. The McBride lineage, because of greater genetic variation, constitutes over 70 percent of the present-day population's genetic ancestry.

Of course in 1972, when the WWF hired McBride, his reputation was not as someone who had played a critical role in salvaging an endangered species from the brink of extinction, but rather as a formidable tracker. If anyone could find out whether any Florida panthers had managed to survive a 500-year onslaught on their existence, it was him.

McBride arrived in Florida with his pack of hounds. He started in Highlands County near Lake Istokpoga and worked his way south over the next several weeks, ending up in Big Cypress National Preserve. He didn't see panthers but he found evidence of some. "Not many," he would say later, "but a few." The next year he repeated the same survey and near Fish Eating Creek, southwest of Lake Okeechobee, his pack treed a panther, an older female, infested with ticks and in poor condition. It appeared that she had never produced kittens.

McBride began teaching Chris Belden, a state biologist, how to spot panther signs such as paw prints, urine markers, and droppings. In 1974, they found evidence that two panthers were living in the Fakahatchee Strand. Based on their searches, McBride and Belden thought that as many as twenty or thirty panthers were still alive, living off deer and feral hogs in the areas around Lake Okeechobee and south into the Everglades. "I was amazed to find them," McBride would say in 1994. "I mean, I got down here in this thickly settled area, and I was really surprised there was any left."

His discovery launched the Florida Panther Recovery Team in 1976. Its task: to come up with a plan to save the Florida panther from extinction.

✳   ✳   ✳

On a foggy winter morning I met Darrell Land on the tarmac of the Naples Municipal Airport in Collier County, Florida. He sported what looked like a fresh haircut, green cargo shorts, and hiking shoes. Of average height with a North Carolina drawl left from his childhood, Land came off as polite but quiet and focused. He pulled a silver laptop out of a backpack and threw the bag in the back of a single-engine Cessna 182P, its tail emblazoned with a blue panther. Once the sun gave off enough heat to dissipate the fog, we would take off with a pilot and head east. Our goal was to search for radio-collared panthers whose signals were received by antennae located under the plane's wings. Land has been flying in Cessnas and listening to these radio signals for thirty years. He was fresh out of graduate school at the University of Florida when the Florida Fish and Wildlife Conservation Commission brought him on board to help monitor and protect the state's panther population. In school, Land had studied cavity nesting birds that make their homes in dead trees on slash pine plantations, but today there are few if any individuals who have more experience than Land in the day-to-day management of the Florida panther.

Our general flight pattern was to troll for radio signals some forty miles north and south of Alligator Alley, the stretch of Interstate 75 that bisects the Everglades. The plane flies three days a week and Land's style of tracking is highly efficient. Once a panther is located on the ground, Land directs the pilot with what are jokingly referred to as "Darrell's patented hand signals" to execute two sharp circles around the cat. Land can ascertain from the radio signal whether the animal is living or dead, information he then inputs on his computer and in some cases relays via cell phone to a team on the ground. There were some thirty panthers outfitted with radio collars in all of Florida, providing researchers with a steady stream of movement and mortality data, and a glimpse into the population as a whole. The job of monitoring ten of them was under the purview of the National Park Service. On a good day, Land and his pilot could locate around twenty panthers over 2,500 square miles and be back at the Naples Airport within two and a half hours.

By nine there was just enough blue sky for us to take off. Nathan Greve, a young, enthusiastic pilot, informed me of the safety protocols: "The two emergency exits are the two doors," he said. With that I crawled into the backseat next to Land's backpack. Built in 1975, the Cessna had the same smell as an old pickup truck: warm vinyl and rust. Greve started the 250-horsepower engine and the propeller roared. As the plane taxied to the runway, we passed by two burrowing owls nonchalantly watching the plane from the ground. The smallest owl in Florida at a mere nine inches, it is also a fast-diminishing species as development eats up its habitat. This pair had somehow managed to create a home in the strip of grass between two runways and spent the days observing private jets and small charter planes take off and land. Greve gunned the motor, gaining speed until we lifted off the ground, the trailer parks and golf courses of Naples spreading beneath us.

Adult panthers have a reddish-brown coat on the ridge of their backs and pale gray underbellies. The biggest and strongest males can reach seven feet long and weigh over 160 pounds. Panthers are lonesome, secretive creatures. They hunt alone and sleep alone. A male and female will spend somewhere between three and five days together under the saw palmetto to mate before going their separate ways. Despite an aversion to humans and preference for thick vegetation, they are surprisingly vulnerable to perils of civilization, particularly roads. It was only the beginning of February and already it was not a good year for the cats. Out of a population of approximately 100 to 150 individuals, two panthers were killed by cars on US 41, one died of unknown causes, and a female perished from complications of giving birth; her litter did not survive. In 2012, automobiles killed eighteen panthers, 12 percent of the total population. In 2014, a record twenty-two panthers would be killed by cars. Since its rediscovery and increased protection, the panther population has grown, but the bigger it gets, the more the cats are in competition with the sprawl of cities and suburbs for space. Each panther maintains an expansive home area. For males, this range measures as much as 250 square miles; for females it is 150. The current population is located on just 3,500 square miles and is in search of new territory; panthers are crossing roads and pressing up against housing developments and farms more than ever before. "There are five times the number of panthers than when I came on in 1985," said Land. "But Florida is a riskier place for them to live today."

In addition to monitoring radio-collared panthers, much of Land's time is spent mitigating the interactions between panthers and humans. To date, there has never been a death or even an attack by a panther on a person in Florida. "We're not a sought-after food item, we're completely off the list," Land explained. But he likes to point to a photograph in his office of a tawny-colored panther sitting assertively next to a birdbath in a backyard. "That's the future of panthers," he said. "Most people love them. They watch Animal Planet and NatGeo on TV and think they're neat and cool. If next to that birdbath there was a sandbox with their three-year-old in it, they would have a different view." Panthers, Land believes, will only succeed if people can tolerate seeing them in their landscape. Recently, this tolerance has been in short supply. Every year for five years in a row, a panther has been shot and killed, a felony offense in Florida. One hunter was fined and sentenced to jail time for shooting a cat with his bow in 2011. "I don't like those damn things," he said. "They are going to hurt someone."

In order for the Florida panther to be delisted as endangered under the Endangered Species Act, there must be two viable populations (defined as one in which there is a 95 percent probability of survival for 100 years) of at least 240 individuals. The goal of the government's recovery plan is to expand the current panther population and habitat north of the Caloosahatchee River, which flows southwest and roughly bifurcates south and central Florida. Panthers have been known to swim rivers a mile wide, and every now and then a male panther navigates the interstate running parallel to the Caloosahatchee and swims across in search of new territory. But there hasn't been a female panther seen north of the river in over thirty years. If a viable population of panthers is going to live outside of southern Florida, it is more than likely they will have to be put there as part of a reintroduction plan. A population of 240 panthers would require as much as 12,000 square miles of habitat; it is nearly impossible to imagine that bringing large predators to the area would be welcomed by the residents, ranchers, or land developers. There are other places that would be possible candidates for the translocation of the cats—Arkansas, or the Florida-Georgia border—places in keeping with their historical habitat. In 2008, a Florida panther made his way to Troup County, Georgia, only to be shot and killed by a deer hunter. As one government official explained to me, "There are no stakeholders that are calling up Florida and asking to bring panthers to their state. It's unfortunate but that's the reality. The belief is they are going to eat all the deer and children at the bus stops." The government agencies responsible for panther conservation have at times behaved in controversial ways that appear to directly compromise their stated conservation goals. In 2010 the *Tampa Bay Times* published a three-part series detailing the "sordid story" of the USFWS ignoring recommendations issued by a panel of experts (including Roy McBride) they themselves assembled, by repeatedly giving a green light to the construction of shopping malls, mines, and suburban development in panther habitats. A key recommendation the agency never pursued was designating a 43-square-mile corridor that would allow panthers to disperse into northern Florida.

Greve leveled the plane's ascent at 500 feet, close enough to the ground to make out the color of pool furniture or watch the arc of a golf ball after tee off. Within a couple of minutes Land picked up a radio signal and went into tracking mode. He directed Greve eastward toward the Picayune Strand State Forest, an area infamous for having been the site of a bold real-estate scam. In the 1980s, the government was forced to buy back land from 17,000 people who had been duped into buying swampland from the Gulf American Land Corporation. The original cypress stands, pine flatwoods, and wet prairie of the landscape had slowly returned but not before the roads had been utilized as runways for South American drug smugglers. "Drug running is a part of Florida's rich history," said Greve.

Land used hand signals to indicate that a panther was near, and Greve slowed the plane to eighty miles per hour. "And we have a cat," he said. Greve banked the Cessna into a forty-degree angle and flew a tight circle over an area of scrubby flatwoods, the open spaces between the pines revealing a blanket of saw palmetto and wire grass. You can spot a panther from 500 feet above the ground but it is easier during the rainy season between June and September, when monsoons force the cats to swim between islands of hardwood hammocks that stay above water. This panther remained hidden and in less than a minute, Land had inputted the location of the animal into his laptop and we were headed east again toward the Florida Panther National Wildlife Refuge and the Fakahatchee Strand Preserve State Park.

Somewhere below us Roy McBride and his dogs had been tracking panthers since dawn. When he first started out looking for the cats in Florida, McBride navigated the sodden, thickly tangled landscape of swamp and forest on foot; today, he and his grandson, Cougar, use swamp buggies and ATVs. McBride once calculated that if a panther moves six or seven miles in a night and its strides are nineteen to twenty-two inches long, it could leave between 19,000 and 38,000 tracks. The evidence the panther unknowingly leaves behind of its existence is what McBride searches for. "We do not go out expecting to actually see panthers," he explained at a conference in 1994. "It is the same procedure wherever we go working with cats, we look for clues. Most are nocturnal. They are secretive. You hardly ever see them." The signs could be a jointed and twisted dropping with hair and bones or a urine marker—a small pile of debris made by the hind legs of a panther, in which they leave a few drops of urine. If vultures are overhead, there is a chance a cat is close by: the birds are attracted to the remains of a panther's last meal. When the hounds tree a cat, McBride calls in a team of biologists and a veterinarian to determine whether conditions are safe enough to tranquilize the animal and conduct a physical exam. The winter months are the preferred time to dart a panther, as the cats are not at as high a risk of overheating. If McBride discovers a den of kittens, each is microchipped, given veterinary care, and a sample is taken for genetic analysis.

These cats are not the same panthers McBride discovered in 1972, at least not genetically speaking. Back then, as McBride, Chris Belden, and a handful of other field researchers spent more and more time tracking and observing the cats, they noticed several unusual characteristics. Unlike in other parts of the United States, the panthers had cowlicks in the fur on the backs of their necks, and a tail with a ninety-degree kink at the end. By the early 1990s, studies showed that 80 percent of male Florida panthers also had cryptorchidism (an undescended testicle) and low sperm quality. Analysis of the population's DNA showed almost no variation between individuals; the cats were nearly genetically identical to one another. In 1994, a group of biologists published a reproductive analysis in the *Journal of Mammology* showing that 94 percent of a male Florida panther's sperm was malformed. All together, these characteristics indicated a level of reduced physical fitness as a result of inbreeding, explaining the high mortality rates among kittens and low reproductive success of the male cats. This discovery was not surprising considering the panther population had been small, isolated, and losing habitat for as many as twenty-five generations. The nearest population of cats was two thousand miles away in Texas, meaning any opportunity to crossbreed was impossible. Population biologists ran computer models that showed the statistical likelihood of the panther population's extinction was near certain within forty years.

Throughout the 1970s and '80s, conservation biologists had recognized that captive populations of animals were demonstrating lower fitness levels due to related parents. Studies of small groups of self-fertilizing plants and experiments introducing new plants had indicated to scientists that one way to deal with the problem was to introduce fresh individuals with different genes to increase variation. In 1990, conservation biologists tried a strategy to create an exchange of genetic material between endangered birds: they switched eggs from nests belonging to two flocks of greater prairie chickens located some forty miles from each other, but the experiment failed. Four years later there were just fifty of the birds left in Illinois and more aggressive measures were taken: 518 prairie chickens were translocated from Minnesota, Kansas, and Nebraska to Illinois and the bird's population began to grow again.

In 1992, a specialist group was assembled at the White Oak Conservation Center, the famed wildlife refuge on the Florida-Georgia border, to discuss genetic management and the panther's future. Among the thirty participants were zoologists and academics, including Stephen O'Brien, head of the Laboratory of Genomic Diversity at the National Cancer Institute, who had studied genetic diversity in cheetahs and lions around the world, and Chris Belden. The group agreed that the population of panthers was demographically and genetically unstable. And they believed that like the case of the greater prairie chicken, genetic augmentation—simulating the addition of fresh genetic material—was the only way to ensure the last remaining population of eastern panthers survived. The group debated various options, including artificial insemination and the release of cats raised in captivity into southern Florida. Ultimately the best route forward, they decided, was to translocate individuals from a wild population of cats and bring them to Florida where their genes could mix with the Florida panthers. This conservation strategy had never been undertaken with official sanction for a species under such scientific and public scrutiny. There was one dissenting opinion among the group: Dave Maehr, head of the field team monitoring panthers for the Florida Game and Freshwater Fish Commission. Maehr would gain a reputation as a controversial ecologist when he went on to consult for Florida land developers; tragically, he died in a plane crash in 2008 doing research on black bears. Maehr's views on panther conservation were flawed but also prophetic. He was convinced that biologists and bureaucrats grossly misunderstood the decrepit state of the Florida panther. They didn't need genetic augmentation; what they needed in his opinion to be healthy and successfully reproductive was suitable habitat and more of it. He wrote a book on the subject in 1997, *The Florida Panther: Life and Death of a Vanishing Carnivore.* The story of the panther's management, Maehr wrote, was a "classic example of what happens when people treat only the symptoms of a much larger problem." Bringing cougars into "the steamy forests of south Florida" was a "quick fix to a complex problem, and soon the Florida panther will be a different animal than the one that has survived all the other attempts to rescue it."

Maehr's concerns were in the minority. As Chris Belden told me, "In 1992, even if the entire state of Florida had been open to panther habitat, we still had the same genetics and the population would've gone extinct." The participants at White Oak issued a recommendation that several panthers from West Texas be introduced to the local population. This subspecies, *Puma concolor stanleyana,* had shared a continuous range with Florida panthers as recently as the 1800s, and the goal was to "reinstate gene flow" lost due to "human-caused isolation." The name of the effort was modified from "genetic augmentation" to "genetic restoration."

Three years later, the best predator hunter in America, Roy McBride, was hired to capture eight females from Texas and bring them to Florida, where they were released into the wild.

✳   ✳   ✳

Competition for funding and publicity in the world of conservation biology is stiff, so it helps if your species can be described with a superlative.

In conservation there is no greater superlative than *rarest*. Guinness World Records awarded the title of "rarest reptile" to the Abingdon Island giant tortoise, of which there was a single individual remaining in the world: Lonesome George. The 100-year-old George was discovered in the Galápagos Islands in 1971, a full sixty years after the last sighting of his kind. Guinness also awarded him the "most endangered species" moniker and he became an icon of conservation and the Galápagos. Biologists tried every imaginable strategy to prevent his genetic lineage from vanishing. Reward money was offered to anyone who could find him a mate. Year after year the searches were in vain. The alternative to total extinction was a different kind of extinction: mate George with another subspecies of giant tortoise and preserve his DNA within the larger genetic pool. Ecologists call it extinction by "anthropogenic hybridization."

Hybridization is a fact of evolution and examples in nature abound. The North American Brewster's warbler is a hybrid of the blue-winged warbler and the golden-winged warbler whose breeding ranges overlap. The sparred owl is a cross between spotted owls and barred owls in the Pacific Northwest. In plants and fish in particular, hybridization is believed to have played a significant role in helping to generate diversity. These insights might not have been possible without the advent of genetic technology. Until then, biologists relied on visible morphological characteristics to detect hybrids in nature. But hybrid individuals do not always display the characteristics of their parents equally, or in ways that are distinguishable to the eye. Once biologists were able to analyze genes, they could "see" the sheer number of living organisms whose DNA was a hybrid of parent species or subspecies.

Genetics does not make hybridization a less complicated matter for conservationists. In fact, the opposite is true. Once hybrids were recognized as ubiquitous, the question of whether they should be protected became controversial. Until the early 1990s, the federal government maintained a rigid though unofficial stance that hybrids between species and subspecies could not be protected under the Endangered Species Act, even if one or both parental populations was listed and no matter if their hybridization

was natural or anthropogenic. In 1991, Stephen O'Brien, the molecular biologist at the National Cancer Institute, and the ecologist Ernst Mayr at
Harvard University wrote an influential article for *Science* challenging the
government's position. "The possibility that a subspecies carries [ecologi-
cally relevant] adaptations coupled with the potential to become a unique
new species are compelling reasons for affording them protection against
extinction," they wrote. According to O'Brien, the USFWS revoked their
"Hybrid Policy" in anticipation of the article's publication, effectively extending protection to at least two controversial species that had been making the news at the time: the red wolf, a hybrid between gray wolves and
coyotes, and the Florida panther, whose inbred stock, scientists were proposing, would improve if they bred with the Texas animals.

Even though the government had a change of heart, it did not offer specific guidelines for how or when to conserve hybrids, and confusion continues to this day. The case of red wolves is a source of debate among biologists.
Are red wolves the product of a hybridization that occurred thousands of
years ago, or just a couple hundred years ago, when hunting and habitat
degradation may have created new behaviors in the animals? If the answer is
thousands, then some believe the species is worthy of protection because it
is a "purer" example of the evolutionary legacy of canids. But if the answer
is hundreds, and the red wolf mated with coyotes because of the human
presence on its landscape, then its genetic stock might not warrant preservation. This distinction between natural and anthropogenic hybridization
can get very murky. For instance, the pallid sturgeon, a dinosaur-like fish
that can live to a hundred years old, is native to the Missouri and lower
Mississippi River drainage basins. In 1990, the fish was listed as endangered
because of human-driven habitat loss and the resulting threat of hybridization with shovelnose sturgeon, a smaller species. But genetic studies have
since shown that the two species are so genetically similar they can hardly
be considered distinct evolutionary lineages; they have probably exchanged
genes throughout their existence. Today, there is an entire population of fish
in the Atchafalaya River in Louisiana that is what scientists call a "hybrid
swarm" of pallid and shovelnose sturgeon. Should shovelnoses therefore be

protected too? Should the hybrid population that contains genes from an endangered fish be excluded from protection? What are we protecting, genes or individuals?

In some instances, the conservation policy has been to prevent species hybridization at any cost. The USFWS in New Mexico euthanized a litter of puppies that they found to be a cross between a rare Mexican gray wolf and a dog in 2011, and then later killed the wolf mother when she was found near dogs again. When biologists discovered that captive Asiatic lions in India had been polluted with African lion genes, it led to the shutdown of breeding programs in Europe and America. In other cases like that of Lonesome George, hybridization is seen as the only strategy for survival. Unfortunately, the last Abingdon Island giant tortoise found his fellow subspecies uninteresting prospective mates, or when he did get excited (sometimes with the helping hand of a herpetologist), the resulting eggs did not hatch. He died on June 24, 2012, and his evolutionary line died with him. In yet another twist, however, researchers announced months after his death that seventeen tortoises from a nearby island shared genetic material with George. Through back breeding, biologists say they might be able to produce a "purebred" facsimile of George within two or three generations, an experiment that could take decades considering the giant tortoise's life span. Whether these offspring would be the "real" thing depends on whether you would describe this sort of human tinkering as natural or unnatural, real or artificial.

Hybridization and the lack of a coherent conservation policy toward it could reflect a deeper sort of ambivalence about how we think about species' identities. Biology tells us that hybrids are a fact of nature and the borders between them are fluid. Hybridization facilitates evolutionary rescue, but we nonetheless think of species in rather rigid parameters; examples to the contrary have long been viewed as transgressions of a natural order. Homer's description in the *Iliad* of the Chimera was of a fire-breathing monster made up of a lion, a snake, and a goat, but most tellingly, it was a "thing of immortal make, not human." Throughout history, chimeras are portrayed as monsters, deities, or angels—not of this world.

The field of stem cell research is now leading us toward an era in which chimeras will appear in our landscape with increasing frequency. Scientists have transplanted human brain cells to mice. They introduced the first transgenic primate, a rhesus monkey born with a genetic code for a green fluorescent protein taken from a jellyfish, calling him "ANDi," which stands for "inserted DNA" spelled backward. Bioethicists have suggested that one reason for our discomfort with these advances, our intuitive revulsion, is that interspecies chimeras and hybrids, especially those created with human genetic material, threaten our unambiguous and privileged status at the top of the natural order. What *is* a mouse with a human brain? The question creates intense moral confusion. Is it human? Are we responsible toward it as we are toward other humans? When we begin to tinker with species evolution to create hybrids and chimeras, we are also tinkering with a moral order.

✳ ✳ ✳

When the eight female panthers from Texas were introduced to the Big Cypress swamps in 1995, the animals did not look much different from the panthers in Florida, which had just a handful of unique characteristics—a darker color than the tawny coat of other panthers, longer legs, and a flatter skull—in addition to the cowlicks and kinked tails from inbreeding. There were plenty of people who questioned whether the two populations qualified as different subspecies at all, including McBride, who said neither he nor his dogs could discern any difference in the cats' behaviors. In 1946, naturalists Stanley Young and Ed Goldman had described some thirty subspecies of cougar throughout the Americas based on morphological characteristics and geography, fifteen of which were in North America. But in 1999, Melody Roelke, a veterinarian, presented evidence at the American Genetic Association based on analysis of 300 mountain lion samples that there were just six genetic subdivisions: five in South America and one for all of North America. Roelke proposed that the fifteen North American subspecies be condensed into one category to be called *Puma concolor cougar.*

The question of mountain lion subspecies remains controversial, but as it was explained to me, whether you believe in separate subspecies or not, Florida and Texas panther populations were former neighbors and there was clearly historic gene flow between the two before human habitation became a geographical barrier.

Nonetheless, the risks of the genetic restoration were twofold. There was a possibility that the new population might suffer from outbreeding depression, a genetic phenomenon in which the crossing between the two different populations produces even lower fitness in offspring. Or the new cats could pose the threat of genomic sweep, in which their fitness so exceeds the original population's that their genes quickly dominate the total genome, effectively driving the original population into genetic extinction. If there was something unique about Florida panthers, it would be lost. Without the genetic restoration, however, biologists estimated that by 2010, there was a 70 percent chance the population would fall below ten individuals.

After the translocation, three of the Texas females died before they could reproduce. The remaining five mated with Florida panther males and produced healthy kittens. Until then, the total number of Florida panthers had fluctuated between roughly nineteen and thirty animals. By 2008 there were an estimated 104. Of the animals that had Texas ancestry, only 7 percent had a kink in their tail, and none that were examined had the cryptorchidism that biologists feared might doom the subspecies. The cats quickly demonstrated a renewed vitality. "Some people describe them as the Arnold Schwarzeneggers of panthers," said Dave Onorato of the USFWS. "Hunting groups are saying they are more aggressive but it's nonsense. The data shows they are more apt to run away now than before. And they are stronger and more vigorous and can do that." There is a small segment of people in Florida, explained Onorato, that still believes panthers are dangerous predators and should not be protected. "We've had these cases of panthers found with bullets in them that are unsolved," he said. "And some people feel now they are really Texas cougars anyway, so they shouldn't be here."

While the restoration undoubtedly extended the life span and health of the population, the intention was to create a population that had no more than 20 percent Texas ancestry in the first generation, a number they hoped would quickly improve fitness and eliminate unfavorable genetic material, the "genetic load," but not represent a genomic sweep. Philip Hedrick, a conservation geneticist at Arizona State University who has worked extensively with wolf populations and has studied the Florida panther case, believes that the percent of Texas stock in Florida panthers today has surpassed 20 percent. "The reason they didn't want to have it be more than that," he explained, "is there might be something unusual or unique in the Florida panther genome to protect. It may have been differentiated or adapted to that environment and 20 percent would allow those adapted variations to remain." So far, no one has analyzed exactly what percentage of the animals in South Florida are made up of Texas stock or Florida stock. One study showed that the animals showed little change, if any, in their unique skull shape after the restoration. And no one knows what these percentages would mean anyway. Would they indicate a new subspecies all together? How much original DNA would an animal be required to have to be called a Florida panther, even if it already hunts, breeds, and sleeps in the cypress stands of South Florida? McBride's patience for these delineations between Texas and Florida panthers is limited. "These people are convinced we've made zebras from a horse and a donkey. We didn't really cross anything. We could call them all cougars or mountain lions," he told me. In fact, he said, when he first started hunting panthers in Texas around Big Bend as a teenager, no one ever called them cougars, just panthers. Biologists recommended in their original plan that new animals from Texas be introduced every five years to continue fortifying the genetic pool, but there is no indication the government plans to introduce more of the cats. One reason may be an intense bureaucratic aversion to the controversy another introduction could create in Florida, where the panthers are viewed with great ambivalence. Unless extinction is imminent again, the political hurdles are too high.

Chances are that genetic "rescues" of this kind will be an increasingly common element of conservation policy in the future. Habitats are becoming more fragmented, not less. In many cases, populations of animals are more isolated from each other, without the porous borders and corridors that could enable a fluid exchange of genetic material and prevent inbreeding. Hedrick cited the case of the Isle Royale wolves, a population that established themselves in the 1940s on an island in Lake Superior by crossing a twenty-mile ice bridge. The wolves survived for decades preying on the island's moose, but in 1980 parvovirus, a disease introduced by domestic dogs, brought the population down to twelve and shrank the genetic pool. In 1997, a lone male wolf crossed the increasingly rare winter ice bridge and became the alpha male of the population. He was virile and territorial, so much so that he displaced one of the island's four packs, driving it into extinction within a couple of years. The wolf, known as No. 93 or "Old Gray Guy," brought a kind of genetic rescue to an ailing population. He spread his genes, producing offspring with higher fitness. But there have not been any new migrant wolves on Isle Royale since. Fifty-six percent of the population's genes are from this one wolf, and the inbred population continues to hover near extinction. The wolf and moose populations on Isle Royale represent one of the longest-running and most closely followed studies of predator-prey relationships in science, and there is intense debate over the right course of action. Should "natural" processes be maintained and the wolves allowed to die out? How much management and intervention is acceptable? Is Isle Royale a wilderness or a laboratory?

*   *   *

The hum of the Cessna's engine and the hot noonday sun lulled me into a state of drowsy reflection somewhere above the Fakahatchee Strand. We floated over bald cypress and royal palm swamps, the light glinting off patches of black water and a sweet smell of baking earth coming through a crack in the window. Darrell Land directed the pilot to circle over another panther, the tiny plane banking sharply gave me a direct view of the ground below. The cat stayed hidden beneath a tangle of thick vegetation. We headed in an easterly direction north of Alligator Alley, which runs all the way to Fort Lauderdale, over the northern swath of Big Cypress National Preserve just below the Big Cypress Seminole Indian Reservation. The landscape was not particularly dramatic, at least not in the way you might describe some other national parks like Yosemite or Yellowstone.

When early American environmentalists such as John Muir presented the rationale for preserving wilderness through national parks, they rarely did so based on the need for biological conservation. Wilderness was valued as a place for people to engage in aesthetic contemplation, solitude, and spiritual rejuvenation. One criterion for identifying national parks, according to the environmental ethicist J. Baird Callicott, was the lands' uselessness for practical purposes; they were too barren or remote for things like farming or industry. The Fakahatchee Strand felt like an extreme example of this early conservation ethic, though it was hard to imagine it as a spiritual refuge. Even from high above, it appeared inhospitable. During the rainy summer months, heat and humidity combine to create temperatures over a hundred degrees in Big Cypress. The swarming mosquitoes are intolerable, and the presence of alligators, snakes, and scorpions unsettling. Our pilot, Nathan Greve, was born and raised in South Florida and spent his youth riding ATVs in the swamps, but he has never gotten used to the brutal summers. "I hate it," he said. Even today, with campgrounds and trails carved into it, Big Cypress and the surrounding Everglades feel like a place hostile to people and not intended for our casual recreation. At the airport in Naples, Larry Harris, another pilot who works with the USFWS, told me the story of how he helped catch a suspected murderer just a few months before. The police in Natchitoches Parish, Louisiana, were searching for a man accused of a triple homicide, but he escaped and headed east to Florida, ditching his rental car in the town of Ochopee before heading into the wilderness on foot. Harris was "flying cats" and saw the suspect walking along US 41 in Big Cypress. He tipped off the sheriff's deputies, who arrested him and sent him back to Louisiana. The man had lasted just a couple of days before the swamp spit him back out.

If we left out a need for biological conservation, what should the rationale be for preserving this landscape today? Relatively few Americans spend any time in the Fakahatchee, nor would they necessarily feel its loss directly should it go the way of Florida's other wilderness areas—under the developer's bulldozer and orange groves. And yet many Americans believe in an ideal of "true wilderness," and in public opinion polls espouse a desire to see more of their federal lands preserved as such. Why do we want to preserve something fewer and fewer of us experience directly?

In 1974, around the same time Roy McBride was treeing some of the first Florida panthers seen in generations, a young philosopher by the name of Mark Sagoff published a paper with the *Yale Law Journal* that presented a compelling perspective on preservation. "On Preserving the Natural Environment" was born of Sagoff's belief that wild places and species must be preserved because they are symbols of American cultural and political traditions. Ethicists continue to discuss his paper because it contributed a critical argument to the field: metaphorical experiences that take place in nature and our national parks do not just fulfill human desires, they *shape* our desires. The discovery of what was viewed as unfettered wilderness, so different from the Old World, inspired Americans' notions of freedom, independence, and self-reliance. The more we lose these wild places and wild species, the rarer these metaphorical experiences become, and the values of the dominant culture shift—toward a different kind of ethic defined by our preferences for consumerism or leisure, say. Extinction, in this context, is a concession to the fact we no longer value the ideals an animal symbolizes. Chris Belden, the only other person who has worked with the panthers as long as Roy McBride, told me that to him the panther is a "wilderness indicator species," meaning that once it is gone, the wilderness in the eastern United States will have disappeared too. He believes that people's interest in preserving this wildness has decreased during his life, in part because so few people even grasp what was here before them. "Most people nowadays are raised in cities or subdivisions; their concept of a wilderness is a state park or a national forest. From a panther perspective, that's not even good range," said Belden. "Up until the 1960s, Big Cypress was impenetrable. People took Model Ts and converted them to swamp buggies, but they could only go as far as the gas would last. Now we have I-75 and all kinds of roads. What used to be impenetrable is now easily accessible."

I kept my eyes on the ground below the plane, trying to glimpse the shiny coat of a panther. I knew the chances I would see one were slim to none. Of the thousands of hours Belden has spent tracking panthers by plane, he said he'd seen one from the air on maybe two occasions. I marveled that the lonesome cats were alive in the brush below us. Their resilience was humbling even as the fragility of their current existence was alarming. From up high, I could see how they are squeezed on all sides by golf courses and airports and interstates. Problematically, the benefits of the genetic restoration for the Florida panthers may be close to expiring. Two of the five Texas panthers contributed a disproportionate amount of offspring, about 70 percent. Even though the genetic rescue increased the population size and fitness of the animals in the short term, there may be a future "bottleneck" awaiting the animals descended from these females. Most significantly, the panthers are still isolated geographically, and therefore genetically, from other populations of panthers. "That fact alone, that they are isolated, means they are steadily losing genetic variation over time," said Dave Onorato. In 2008, Roy McBride published a paper on his methods for estimating puma populations in Florida and noted that today's panthers could be reaching the carrying capacity of their current environment. Depredations of cattle are on the rise, automobile mortalities too. An introduction of more panthers from Texas might once again increase the population size, but that would only exacerbate the underlying problem: there is no place for a viable population of panthers in the southeastern United States. "We can't just crate up a bunch of panthers and take them to Arkansas and let them go," said Land. "It's been hundreds of years since [people in Arkansas] had to live with panthers. I'm not optimistic that reestablishing panthers outside of Florida is going to happen."

The only natural mechanisms for Florida panthers to reestablish genetic diversity is to mutate or migrate, and so with their limited genetic stock, the future of the species may depend on whether panthers can freely reestablish parts of their former range on their own. This would give conservationists and government agencies a free pass to avoid the political, social, and legal hurdles of roping off new critical habitat for the species. If a few female panthers would only venture north and give birth to some litters of kittens,

I was told, it would be the best thing to happen for the subspecies in forty years. For this reason, the Caloosahatchee River, the northern border of the panthers' current territory, has become a literal and symbolic crossing in the Florida panther story. There has not been a female sighted there in over three decades. Without them, the few male cats who swim across are just stragglers and wanderers.

The one person I felt sure could offer insight into the behavior of the animals and therefore the future of the species proved elusive. Save for in a few academic papers and articles in small magazines, Roy McBride has gone on the record just a handful of times with thoughts on his role over fifty years in the fates of the Mexican gray wolves and Florida panthers. Based on these comments, he continues to stand out as an unlikely character among conservationists and government policy makers. "I don't think ranchers should be imposed upon by someone else's will who lives way off in a city or somewhere else," he told *Ranch Magazine* in 1984. "At the same time, it is not the duty of the taxpayer to kill the rancher's coyotes. It is the problem of the man and he should take care of it but he shouldn't be interfered with. That's how I see it—it's real simple." McBride was once asked if he considered himself a conservationist and he reportedly laughed and replied, "Hell no." But there are intimations in his public statements of real regret over the disappearance of large predators from our landscape, even if he once helped to extinguish them. "They didn't leave a heritage," he told *Texas Parks and Wildlife* in 2012 about the disappearance of red wolves in Texas. "They didn't leave a building you could look at or dig a big hole or put in a dam. I guess the first rain that came along after the last one was caught washed out his tracks, and that was about the only sign they were there. We'll never have them again."

In his book *The Ninemile Wolves,* Rick Bass described seeing McBride speak to a crowded auditorium at the 1990 Arizona Wolf Symposium.

> "I've done it all," McBride says calmly. . . . He speaks without a text, holding his big hat in his hands. McBride is lean, flat-bellied, square-jawed, tall and boyish looking. We can't recognize him as prey, or enemy—he

doesn't fit our image. McBride looks out across at us and continues his modest yet accurate understatements. "I've done some work in Mexico tracking wolves," he says. "It's kind of sad that there's this many people around interested in them and not any of 'em left around to work with. I think I had the best job anybody ever had," McBride says. . . . "It was worth it to get to see those tracks and the things they did. I had no idea we could ever get rid of them."

Land told me he thinks McBride's part in the conservation of Florida panthers is connected to his work starting out as an animal depredation guy. "At a point you develop a certain respect for the animals, and maybe this was a way for him to pay back a debt he felt he owed to the species," he said.

One evening in late winter I got a voice mail from McBride. For about six months, I'd tried to reach him but heard nothing back. Part of the difficulty, I found out, was that McBride doesn't use a computer. Someone else prints his e-mails out and physically delivers them to him each week. He formulates a response and the same person takes them back home, types them up, and sends them out. "It's still slower than snails," he said in a strong Texan cadence. But the real reason it took so long, I soon figured out, is that McBride eschews attention for his work. "I was just the first person to catch one," he explained. "I wouldn't want to toot my horn about that." Over the next year, I kept in touch with McBride, even meeting him at his home near Ochopee to share what I had written about panthers. "I didn't do something that someone else couldn't have done, I just happened to be around when something was happening with them," he told me. "It's just history and I can't change that." Unfailingly gracious with a humorous spark, McBride kept his opinions about the conservation effort close, offering only that it had often been contentious and political. "It hasn't always gone real well," he said. "It hasn't been all kittens and rainbows." People made whole careers out of trying to save panthers, but few had spent any time with the animals or gotten to "mess" with them. "It takes me about three sentences to figure out if they have any experience with them," he said. The other problem, as he saw it, was that people involved in the effort all have different versions of the story. "It's like seeing a car wreck and everyone saw it different. This is just like that. Some of it is downright fraudulent."

McBride had suggested that I talk to his son, Rocky. This took a little while as the younger McBride was in Brazil, capturing jaguars for a population study. Rocky has captured snow leopards in Mongolia and Kazakhstan, Siberian lynx in the Soviet Far East, and pumas all over North and South America. But jaguars are his love. Some twenty years ago, he bought over 120,000 acres of land in Paraguay where he could bring clients to hunt the sleek cats. A year later, the country joined the Convention on International Trade in Endangered Species (CITES) making exporting any jaguar trophies out of the country impossible. At the same time, development and cattle ranching in the country boomed. Around Rocky's ranch as much as 80 percent of the forest was being cleared and turned into cattle pasture, and bounties were placed on dead jaguars. Rocky only had to look as far as the southeastern United States to see what was in store for the cats unless something was done. He began focusing on conservation strategies, working with private landowners and the government. Rocky believes the cause of habitat loss is simply economics, and only economic incentives will conserve habitat. "For a South American to lose ten cows to a jaguar, it's like a disease," said Rocky. "If there is no component or incentive to protect live jaguars, or any large cat that needs large land requirements to succeed, they are going to be reduced to parks and such." Total protection of the jaguars, said Rocky, outlawing any killing, is doomed to fail because the laws are unenforceable in Paraguay. His vision is for a national program of sustainable use—regulated hunting—much like models of game conservation in South Africa. "I think the greatest conservationists are the people who hunt," he said. "There's a lot of money out there, and some of the NGOs say protect this and save that and we have a crisis and generate a lot of funding. They don't want to solve a crisis, they want to maintain it."

These types of politics are exactly what Rocky said his dad wants to avoid when it comes to Florida panthers. "What he does is capture the cats," he said. Rocky described how his dad got a job with the government's predator control agency before he was out of college with a degree in wildlife biology, taking a spot as a hunter in Texas and how, from as early as he can remember, Rocky was along for the job, sitting behind his father on his horse or a mule. He was even with his dad during the hunt for Las Margaritas. "I remember seeing its tracks, headed up this road toward a trap. Something happened and it missed the trap," he said. "It was a challenge. He's always liked challenges." When Rocky was seventeen, his dad sent him to Florida to look for panthers; now Rocky's own son, Cougar, works full time tracking panthers in Florida, an ongoing lineage in a field that is otherwise disappearing. "It's a very small niche," he said.

Over the decades the cats have become sturdier, but Rocky put their future in bleak terms: "Central Florida is developed orange groves and Disney World and everything else. There is no real habitat. There is some in the northern part of the state and southern Georgia, but short of man moving the cats, there's just no way for a dispersal. It gets more difficult all of the time."

EXUBERANT EVOLUTION
IN A DESERT FISH
Cyprinodon tularosa

n the Chihuahua desert in southern New Mexico, I walked along the banks of a small creek following the tracks of a coyote that had trotted this way during the night. The animal's feet had sunk into the sulfurous mud and the ridges in between each of its pad prints were dusted white. At first I had assumed this was gypsum, blown in from the nearby dunes of White Sands National Monument, but it was actually salt, leeching from the ground and eventually washing into the water of the creek. Called the Lost River, this stream is hypersaline, meaning its water is several times saltier than seawater. In parts of its mile-long course, the water's salt content has been recorded as high as 100 parts per thousand (the ocean is on average 35). And yet as the sun cast my shadow over the water, it disturbed schools of tiny fish that flitted this way and that. Their presence in this desert landscape was a marvel. Somehow they survived in a habitat that biologists would otherwise consider lethally toxic to fish. Furthermore, they were a freshwater species. Called the White Sands pupfish or *Cyprinodon tularosa*, they have helped change our understanding of what evolution is capable of and the degree to which humans can influence it.

To understand how, we have to start with a question that has bewildered and preoccupied scientists for hundreds of years. What is a species? Scientists tell us that species are the principal units of evolution. But if evolution is a process of changes in the genetic makeup of a population over generations, how can we define species? This riddle is called "the species problem," and it has everything to do with how we think about nature and the exuberant process called evolution that has produced so many billions of life-forms. Before the nineteenth century, people believed species were divinely created with fixed identities: an animal had an essence that gave it membership in a group with other animals with the same essence. Carl Linnaeus, the Swedish botanist who invented the species classification system, believed that he was cataloging earth's creatures as created by God. After *On the Origin of Species* was published in 1859, it was impossible to rationally refute that species actually changed over time. This made defining a species confusing. Where do they start and end? Darwin himself recognized this problem when he wrote that "the domestic races of many animals and plants have been ranked by some competent judges as the descendants of aboriginally distinct species, and by other competent judges as mere varieties."

Understanding the process that created species is intrinsic to understanding what a species is. But this quest has led brilliant minds into scientific and philosophical sinkholes for the last 150 years. And for all our knowledge about biology today, the issue has never been definitively answered. Today there are about twenty-six different "species concepts." The most simple of these is the biological concept, first developed in the 1940s by the renowned evolutionary biologist Ernst Mayr. Mayr defined species according to their potential for reproducing. He said that species are a group of natural occurring populations that are reproductively isolated from other groups; they can't interbreed or produce fertile hybrids. A horse and a giraffe, for example, are different species according to the biological concept, because they can't procreate. This definition has been described as problematically basic. Scientists have discovered a lot of examples of animals that they can't prove exist according to fixed reproductive rules. Some asexual animals, like starfish and sea anemones, exempt themselves from the mechanics of breeding outright. But we wouldn't say that the hundreds of species of starfish aren't species, even if they don't fulfill the biological concept's specifications. Another concept, the phenotypic, defines a species based on shared physical characteristics. But appearances can be deceiving and unhelpful. What degree of difference determines a new species? Differences between populations or individuals might be irregular, the product of environment or what's called genetic drift, the random disappearance of genes as individuals die or don't reproduce. The colorful plumage of a male mallard duck looks different from the somber female mallard duck. They're the same species, but using the phenotypic concept to define why is unhelpful.

In the early 1980s, biologists came up with a new species concept, something they felt better encompassed life in all its abundant variation. Called the phylogenetic concept, it says that species are the smallest cluster of organisms that share a common ancestor. Since the 1980s, molecular genetic analysis has given scientists a very precise tool for determining common ancestry. By focusing on specific markers in an animal's genotype and comparing it to a related animal, they can determine complex demographic histories including how long ago the animals split from one another into separate branches on the evolutionary tree. The problem is that the phylogenetic concept often ends up splitting species into very small groups, and this can dramatically balloon the number of species we understand to exist on earth. (In a few unique cases, the phylogenetic definition reduces the number of species within a taxonomic group: the number of deep sea snails shrinks from two to one, and mollusks as a whole decrease by 50 percent.) Consider, for example, that the number of lichens in the genus *Niebla* goes from eighteen to seventy-one species. Birds of paradise in New Guinea increase from some forty species to as many as 120.

Adherents to the biological concept call proponents of the phylogenetic concept "splitters," and in turn, biological conceptors are known as "lumpers." As the evolutionary biologist Jody Hey has written, the conflict between these two groups comes down to whether small differences between organisms don't really matter, or whether small differences are the very stuff that makes up a distinct species. It may seem like a turgid theoretical debate, but it has big consequences for how we intervene to conserve animals. In 2004, researchers published a study in the *Quarterly Review of Biology* analyzing over 1,200 species previously defined according to Mayr's biological concept, and reconsidered them according to the phylogenetic concept. The result was that the number of species increased by 48 percent. Even in well-studied groups such as mammals, arthropods, and birds, the authors reported an increase in the number of species as high as 75 percent. Applying a phylogenetic concept also changed the number of species potentially designated as vulnerable and endangered, and they estimated the cost of protecting these "new" species at about $3 billion in the United States alone.

This figure is why politicians and policy makers might have strong opinions about lumping or splitting. The taxonomic classification of a species can change an animal's status under international treaties or the Endangered Species Act. In 1978 the United States Congress weighed in on the debate, ultimately rejecting an attempt to adopt a strict biological concept of species in its application of the Endangered Species Act, which had been passed five years earlier. Instead, lawmakers changed the language of the law to include "species," "subspecies," and "distinct population segments." This last term is somewhat befuddling. How distinct does a population need to be to warrant protection under the law? In the mid-1990s, policy makers defined "distinct" to mean an "evolutionarily significant unit" that is part of the legacy of the species. ESUs, as biologists call them, are populations that are basically on a path to speciation; they are genetically differentiated enough to warrant protection and their own management plans. Trying to decide what makes an ESU is dubious business, but they can be useful for conservation purposes. Consider the case of the Puritan tiger beetle, which lives in two places, the Connecticut River and Chesapeake Bay. Because they are designated as ESUs, no one can make an argument to disturb their habitats just because another population exists elsewhere.

Of all the species concepts, there is one that I found the most satisfying and intuitive. The paleontologist George Gaylord Simpson, considered one of the great scientists of the twentieth century, developed it. Simpson worked at the American Museum of Natural History in New York City for thirty years studying fossils, and he believed that species are "lineages of ancestral descent" that maintain a separate identity from other lineages; each has its own evolutionary tendencies and historical fate. Species are, in other words, organisms that share a trajectory through time. The ways and means of this trajectory consumed much of Simpson's brilliant mind. In the late 1930s, when he was a young and rising star in paleontology, he began focusing his energy on developing a theory that would help explain how evolution works to create species by merging the disparate fields of paleontology and genetics. His unique and profound insight was that evolution takes place at different speeds.

In 1944, after a delay caused by serving in World War II, Simpson published his seminal book *Tempo and Mode in Evolution*. It arrived during a period when paleontologists disagreed as to whether natural selection was the crucial mechanism driving evolution. The problem is that the fossil record is an imperfect one; it contains very little evidence of natural selection producing organisms in transitional states of evolution. If, as Darwin proposed, every new species descends from a preexisting species, where was the evidence of the intermediate stages in a species' development? In 1995, American biologist Niles Eldredge wrote that for paleontologists evolution never seemed to happen: "Assiduous collecting up cliff faces yields zigzags, minor oscillations, and the very occasional slight accumulation of change over millions of years, at a rate too slow to really account for all the prodigious change that has occurred in evolutionary history." Darwin himself felt the lack of these links in the fossil record was the "most obvious and gravest objection which can be urged against my theory."

Simpson, however, believed that the glaring absences in the fossil record didn't refute natural selection; they were due to "quantum evolution," when transitional forms of organisms either rapidly leaped into higher taxonomic groups or vanished into extinction. Until then, scientists had assumed that evolution worked at a truly glacial pace. It was, as Darwin said, happening at every moment "rejecting that which is bad." But, he continued, "we see nothing of these slow changes in progress, until the hand of time has marked the long lapse of ages." Simpson disagreed. Yes, evolution was so slow at times it appeared nonexistent, but at other times quantum leaps in speciation were taking place that were so fast they eluded the stratigraphic record all together. In addition to this quantum speed, Simpson proposed three evolutionary tempos: slow, medium, and rapid. How fast evolution takes place, he said, is the result of variables such as genetic diversity, mutations, generation length, population size, and natural selection, all operating on the gene pool of a population of organisms.

This notion that evolution could work at various tempos provided a foundation for other scientists to investigate its potential power and speed. One of the first people to do this was Scottish scientist J. B. S. ("Jack") Haldane, who created a unit of evolutionary measurement a few years after Simpson published *Tempo and Mode*. Haldane's invention, what he called a "darwin," was a tool that scientists could use to calculate the rate of evolution in a species. To come up with the darwin, scientists essentially subtract a trait value of a particular time or population from another trait value of a time or population, and divide the resulting value by length of time in millions of years. With this formula, scientists could estimate the rate of evolution of the triceratops, for example, at 0.06 darwins, or 6 percent per million years. This million-year time frame was considered appropriate because Haldane believed that even the fastest pace of evolution took many eons to occur.

This was why I was standing on the banks of the Lost River looking at pupfish. I had been told that they were proof that Darwin, and even Simpson and Haldane, had been mistaken.

✳   ✳   ✳

The White Sands pupfish could be the only species whose extinction threats include accidental missile strikes. Of the fish's four habitats, one is on a US

Air Force base and the other three are on a US Department of Defense missile range. The presence of the fish on land managed by the military has had a two-fold effect on the species. On the one hand, the highly secure boundary delineating tens of thousands of acres as off-limits to public or private use has protected the fish from environmental interferences like water extraction or groundwater pollution from farming and other development. On the other hand, not a lot of people get to study them. "They've been ignored except for a few sporadic studies. The military doesn't like the public out there and even being there is dangerous, there's all kinds of exploded stuff everywhere," said Craig Stockwell, a professor of biology at North Dakota State University. As I stood on a steep bluff overlooking the Lost River to survey the area, I had a chance to see for myself what he meant. Just half a mile to the west was the world's largest test track, a ten-mile tarmac strip where military engineers attach equipment to a rocket-propelled sled to see how they will hold up during flights. And as I made my way through the sagebrush, I accidentally kicked a pile of spent rifle ammunition with my boot.

Stockwell, an ichthyologist, has been studying White Sands pupfish since the early 1990s, when there was talk about raising their federal threat level from "species of concern" to endangered. They were already listed as threatened by the state of New Mexico, but a higher federal listing would have been extremely troublesome to the military. The perimeter of protection afforded to their habitat would likely have expanded and changed how close weapons-testing and other activities could take place. In the 1990s, the Clinton administration often sought to avoid these kinds of conflicts by coming up with agreement plans, in which all the stakeholders would come to an arrangement on how to best protect a species while not severely limiting activities or development.

In 1994, the military and other wildlife agencies came up with an agreement for the White Sands pupfish. It called for annual monitoring of the fish population as well as further scientific research. Because so few people had actually studied the species, there were a couple of peculiar mysteries surrounding its natural history. When geologists first discovered them in the early twentieth century, they had found fish in just two locations, a highly saline stream called Salt Creek and a freshwater wetland area known as Malpais Spring. By the time the pupfish was formally described in the *Southwestern Naturalist* in 1973, however, the fish were also found in two other locations: Lost River and a drainage area to the north called Mound Spring. The question was whether these populations were all worthy of protection. Did they represent evolutionary significant units or were they genetically similar? To know this, biologists had to find out how exactly the fish had arrived at these four places.

In 1995, Stockwell collected pupfish from all four populations as part of a $200,000 government grant to look at the genetic structure and life history of the species. Comparing their DNA, he began compiling a history and discovered that the Malpais Spring population contained the highest diversity. These fish had likely split from the Salt Creek population less than 100,000 years ago, most likely in the late Pleistocene epoch (later he would pinpoint it to around 5,000 years ago). The separation between the two populations as indicated by their genetic divergence made sense to Stockwell, especially because he could see with his own eyes that the two populations looked morphologically different from one another. Saltwater and freshwater have different densities that result in different shapes of fish. Saline habitats produce fish with more slender bodies and hence less drag that allows them to move fast, whereas fish in freshwater have a deep-bodied shape. Based on his genetic analysis coupled with these observations, Stockwell recommended that these two populations of pupfish be designated as evolutionary significant units warranting equal protection. "To have an ESU is a pretty stringent requirement—if you have that, you have different species," explained Stockwell. Meanwhile, he found that the other two populations of pupfish, Lost River and Mound Spring, were genetic derivatives of one population; they had originated in Salt Creek but somehow were now found in two new locations. How this came to be required some sleuthing.

Around the same time Stockwell was analyzing pupfish for the genetic study, an ecologist by the name of John Pittenger was searching for clues in the historical record about the species. Pittenger had been involved in White Sands pupfish management since 1994 when the New Mexico Department of Game and Fish realized that the population of feral horses in the Tularosa Basin—a holdout from the ranching era—were drinking from the freshwater at Malpais and Mound Springs and endangering the fish. Pittenger had advised a roundup (today there are just a couple of horses left) to move the horses out. Now he was a contractor for the government and part of the pupfish management team. Pittenger was in a sense working toward the same goal as Stockwell, establishing a timeline for the arrival of the fish in Lost River and Mound Spring. He began by asking questions of the local community and digging into archival records. At the Museum of Southwestern Biology in Albuquerque, he found the papers of William Jacob Koster, an early New Mexican ichthyologist, and looked for references to the pupfish. "I was going through notecards and I found this little index card that had a note about an R. Charles and the Salt Creek fish," Pittenger told me. "So I found this guy's son out in San Francisco and he said 'I got all my dad's stuff, let me look.'" Among Ralph Charles's personal letters were a bunch from the 1960s requesting security clearance from the missile range to visit the pupfish located at Salt Creek. Pupfish fascinated Charles, a retired employee of the Bureau of Reclamation, a water management agency under the US Department of the Interior, but he was repeatedly denied access by the military. Finally he appealed to a US senator, who granted him a one-day pass onto the missile range. On September 29, 1970, Charles visited the Salt Creek pupfish and then did something bizarre for reasons that have been lost to time: he took thirty of them and brought them onto the White Sands National Monument, where he walked into the gypsum dunes and released them at the bottom of the Lost River. Later, he would write to the manager of the White Sands National Monument asking how the pupfish were faring.

Moving fish is nothing unusual. Throughout the 1960s and '70s, conservationists had been using translocation to deal with threats to vulnerable populations of fish, and this was particularly the case when it came to desert species. The presence of these organisms in such an arid landscape says a lot about their tenacity and adaptive powers. Pupfish are found in some of the most hellish, stressful environments imaginable for a living organism. Of the fifty species of pupfish that diverged from a common ancestor millions of years ago, thirty are found in the southwestern United States, remnants of a geological era before desertification when lakes and rivers dried up and left them isolated in at times incredibly small habitats. *Cyprinodon macularius,* for instance, known as the desert pupfish, lives in drainages in Baja and Sonora in water temperatures that range from 40 to 113 degrees. *Cyprinodon diabolis,* the Devil's Hole pupfish of the Mojave Desert, has survived for as many as 20,000 years in a single cavern fed by an aquifer, where the water's oxygen levels are so low they would be lethal to any other fish. (In recent years the population has dropped from a high of about 500 to 35 for reasons biologists don't understand.) "I like to say they are evolutionarily labile," a biologist told me about pupfish. "They are capable of evolving in multiple directions. They can be saline fish, freshwater fish. These changes would normally cause extinctions in most fishes but they tend to persevere." However there's a limit to their adaptability; they can't migrate. So if a farmer in the Southwest decides to divert a stream, or a land developer wants to build a parking lot, a pupfish can't get out of the way. At these times conservationists would step in, often split a population and introduce one to a new habitat, or reestablish a species to a historic habitat. In the 1980s, translocations were so popular as a conservation strategy that more than 80 percent of recovery plans for endangered and threatened fish called for them.

Of course Ralph Charles wasn't a conservationist. His motives for moving the fish were murky: perhaps irked by the difficulty in visiting the fish on military land, he moved them to White Sands National Monument where he could visit whenever he wanted. For Pittenger, the last remaining mystery was how the fish also ended up at Mound Spring, a place that didn't appear to be part of any historic habitat range of the pupfish. After extensive interviews he discovered that the spring had been excavated in 1967, and the resulting pond was likely stocked with fish from Salt Creek around then, maybe as a form of mosquito control.

If it hadn't been for these two haphazard translocations, no one may have discovered the evolutionary significance of the White Sands pupfish. It was Stockwell who sensed that over the previous three decades, while no one was watching, a perfect experiment had inadvertently been conducted. Because while the Lost River was very saline, Mound Spring didn't have much at all; it is a freshwater habitat. "It was those two pieces of information that allow the populations to be compared," said Stockwell. His hunch was that the pupfish might have something to reveal about rapid evolution, a subject he had been interested in since reading a paper in 1990 by respected biologist David Reznick from the University of California, Riverside. Reznick had been studying guppies in the Caribbean since the 1970s and his paper "Experimentally Induced Life-History Evolution in a Natural Population" presented remarkable findings based on years of field research (it has since been cited over 700 times). Reznick had been looking at a population of guppies in Trinidad whose main predators preferred large, sexually mature guppies. As a result, the guppies appeared to mature at an earlier age in order to reproduce faster, and also produced smaller offspring than other guppies. Reznick decided to move some of these fish to a location where the predators focused on smaller, juvenile fish. Within a couple of years after being translocated, the guppies were maturing later and producing less young, changes Reznick showed to be heritable in subsequent generations and therefore genetically based. This was direct experimental field evidence that a selective pressure such as predation molded evolution in guppies over a short time span, in this case around thirty to sixty generations. But the longest of Reznick's field studies had lasted eleven years, and he pointed out the need for longer-term data. Now Stockwell had access to a field-experiment that was thirty years old, begun way before most biologists had the tools or foresight to guess that the process of evolution might be taking place much faster than anyone—even George Gaylord Simpson, the first paleontologist to understand that evolution worked at different speeds—had been able to infer from the fossil record.

In the summer of 1996, Stockwell set up what's called a common garden study, a controlled experiment in which organisms are moved to new environments to see how they produce offspring over time. He set up thirty-six plastic pools with gravel and artificial grass and stocked each one with ten male and ten female pupfish. Some pools were only stocked with fish from Salt Creek and others from Malpais Spring, some had a combination of both, and half of the pools had low saline levels while others had high saline. After the first generation of fish was born and grew to reproductive size, Stockwell sacrificed them in ice water and preserved them in jars of ethanol with the intention of comparing all of them. Then in 1998, his funding for White Sands pupfish studies came to an end and he went to teach at North Dakota State University. He brought the jars with him but they stayed sealed for years, until an eager graduate student named Mike Collyer came along.

※   ※   ※

It was Mike Collyer who had advised me to go to Lost River to see the pupfish. "It's just amazing," he told me. "You get the gypsum dunes in the background and the Lost River disappears into them." When I arrived at Holloman Air Force Base, a young lieutenant drove us out to a bluff that overlooked the Lost River, and I saw what Collyer had described. The "river," at times no more than a foot wide, crept through the arid earth speckled with blooming golden crownbeard and creosote bushes until it appeared to be swallowed up by the gypsum dunes. Some of the dunes can reach 30 feet high, and they cover hundreds of square miles of the Tularosa Basin. Walking over them feels lunar. The forces that produced the dunes started around 250 million years ago, when this entire region was underwater. As the waters of the shallow ocean receded, they left behind gypsum minerals that became part of the stone. Some 70 million years ago, the formation of the Rocky Mountains, an event known as the Laramide orogeny, lifted the floor into a giant arch that then collapsed 60 million years later to create a depression in the earth. As the gypsum leeched from the rocky escarpments of the San Andres Mountains to the west and the Sacramento Mountains to the east, it was swept down by rain and snow into lakes in the bottom of this basin. And when these lake waters evaporated, the gypsum was exposed and great crystal beds of selenite formed. Eventually, the ravages of time broke the crystal into smaller and smaller particles, so tiny the wind picked them up and built the white dunes that grow and shift like torpid waves across the basin floor today.

The Lost River doesn't take its name from this visual effect of dissolving into the dunes, but most likely from a nineteenth-century geologist who wrote in *Science* about the "lost rivers" of the Tularosa Basin, which he guessed to be a giant former riverbed. Indeed, the ancestral Rio Grande once flowed through the basin on its way to the Gulf of Mexico, and this likely explains how pupfish arrived 2 million years ago: by swimming upstream through the original lost river. Around 10,000 years ago, at a time when sloths, camels, and mastodons were roaming the southwest, the climate began to warm and the land became increasingly arid, turning coniferous forest into desert and drying up the lake where the pupfish had survived. As a result, the fish were isolated to tributaries on the outskirts of the basin. In 1885, a biologist reported Native American origin stories that described the basin as undergoing "a year of fire" during which the "valley was filled with flames and poisonous gases." Geologists now know that a volcano called Little Black Peak on the northern border of the basin erupted 5,000 years ago, spewing black lava that spilled into the basin and created a new freshwater spring, called Malpais, meaning "bad country" in Spanish, where the pupfish established themselves. As the lake continued to shrink, the pupfish were isolated to these two places, Malpais and Salt Creek.

From the middle of the basin, which covers some 6,500 square miles, the surrounding mountains are muted blue on the horizon. Mike Collyer originally came here as a master's student under Craig Stockwell. "The fact he was working on desert fish, that statement was so cool to me," said Collyer. "I was a student looking for a decent graduate experience and I had no concept of where it would go or how it would develop at that point. I thought pupfish were really cool fish: the idea they could survive freshwater or two to three times the salinity of the ocean. From a biological standpoint, I found them fascinating." Collyer began doing surveys at Mound Spring, where the fish had recently undergone a population crash that he and Stockwell suspected was caused by parasites, exacerbated perhaps because the translocated fish were less immune to them. It was tricky for Collyer to find the parasites because they were about the size of a grain of sand. (Stockwell's initial discovery of these gastropods inadvertently led to the creation of a new genus of snail, which they called *Juturnia* after a Greek god of springs.) As Collyer conducted his surveys, he became distracted by something else. The Mound Spring pupfish didn't look the same as the fish at Lost River. "You spend enough time looking at the fish and start thinking, you know, they look different to me," he said. "It's a lot like parents of twins, they see differences in their kids other people can't see."

What started as a tangent became the main thrust of Collyer's research. He wanted to find out whether the population at Mound Spring was actually morphologically different from the other populations, so Stockwell sent him to South Carolina to study alongside James Novak, a biologist with an expertise in what's called ecological genetics, how organisms interact with their environment and evolve in size and shape. Together they used an analytical technique developed in the 1990s called geometric morphometrics, in which "landmarks" or points are plotted on a photograph of an organism to create shapes that can then be compared to one another. In 2001, Collyer laid traps in the springs, capturing nearly 400 pupfish. He brought them to a laboratory on Holloman Air Force Base where he labeled and photographed each one. He plotted thirteen landmarks including points on the eyes, tail, and dorsal fins, and then began analyzing the shape. Collyer quickly found that the Lost River population had maintained a streamlined shape, but the Mound Spring population, in just three decades, had evolved a deep body shape more pronounced than even the native population at Malpais Spring. In his resulting paper with Novak and Stockwell, he reported, "Because Lost River and Mound Spring populations could potentially serve as refuge populations of the native Salt Creek population, it is important to evaluate how novel environments could affect the evolutionary legacy of the Salt Creek [evolutionary significant units] in refuge environments."

The paper became the first chapter of Collyer's doctoral dissertation, but there was still a crucial, unanswered question. His study had shown differences in morphology but it didn't explain why. Was it just the result of phenotypic plasticity—the ability of an organism to change its appearance in response to changes in the environment—or an example of something much more unique to observe, changes to the pupfishes' DNA? Collyer suspected the latter and if he was right, and the changes in the pupfish were genetically based, it would represent a rare example of recorded contemporary evolution—defined as heritable changes within a few hundred generations—in the wild. To find out Collyer used the jars of preserved pupfish that Stockwell had gathered in 1998 for his controlled experiment with the pupfish. Using the same geometric morphometric analysis tools, Collyer discovered that the first generation of fish born in the common garden study showed similar shape differences to the pupfish in the wild. The deep-bodied shape of the Mound Spring population was not a "plastic" response to lower salinity, nor the result of random genetic drift, but a case of evolutionary divergence from their original population. "This was contemporary evolution," said Collyer. "This population of fish was now different from its source. That was fascinating. We're talking about a couple of decades, when we're used to seeing changes over thousands of generations."

When it came to measuring the rate of evolution of the White Sands pupfish, Collyer and Stockwell didn't use darwins as their unit of measurement. Instead they used something called "haldanes," a more recent and agile unit of measurement. Haldanes were developed in 1993 by paleontologist Philip Gingerich, who wanted to create a measurement that allowed scientists to compare rates of evolution in different species, or in the same species but on different time scales. Darwins weren't up to the task because they only allowed scientists to measure change per million years and couldn't give an accurate estimate of the intensity of selection taking place. To do this, Gingerich decided to calibrate evolutionary rates over generations instead of years. The definition of one haldane is the change by a factor of one standard deviation (sd) per generation. The standard deviation might be the weight of a New Zealand chinook egg, for example, and the change of weight expressed in haldanes might be expressed as 0.048 sd/generation. This system allowed scientists to create an evolutionary rate for anything they wanted to measure—for instance, the diameter of an eye or the length of a bird's beak. The measurement could then be put in context to get a truer sense of the speed of change. "It allows us to compare the evolution of a whale flipper to the wing of a fly," said Collyer. "The units are standardized and we can say whether the rate is small or large. As soon as we could compare different rates for different taxa, we had a tool for asking, 'Is what I'm looking at impressive?'"

With the White Sands pupfish, the answer was yes, it was very impressive. For females the body shape differences in wild populations were 0.174 sd/generation and 0.159 sd/generation for males, one of the fastest rates recorded in a vertebrate species. The numbers expanded what biologists thought possible when it came to rates of evolution. In their 2011 paper, Collyer and Stockwell wrote that they recognized the contemporary divergence of introduced populations might be used as a tool to enhance biodiversity in pupfish. But if the non-native populations were no longer genetic replicates of the original populations, reintroducing them and risking that they were now maladapted "might be too costly a gamble." Instead, they warned, refuge populations like the pupfish at Mound Spring "might best be viewed as evolutionary experiments."

For decades, conservationists had been translocating fish and other species or creating captive breeding programs to mitigate extinction threats all over the globe. Now it was likely that these actions were capable of setting a species on new evolutionary paths, just like Ralph Charles and an unknown farmer had inadvertently done when they introduced pupfish to the Lost River and Mound Spring. And these evolutionary trajectories are no longer "wild" in the way we typically understand that word to mean free of human interference. George Gaylord Simpson defined the concept of a species as ancient lineages with their own evolutionary tendencies and historical fate, but he most likely didn't anticipate that humans could influence this fate in such a direct fashion. It's not that no one knew humans could cause extinctions—in the early twentieth century there were numerous contemporary cases to point to, from dodo birds to Tasmanian tigers. But now it's clear that the same forces of environmental change that result in extinctions are also speeding up evolution. And in some cases, the rate of environmental change can exceed the capacity of populations to adapt. "It makes intuitive sense," explained Stockwell. "The evil quartet of extinction—invasive species, overharvesting, habitat destruction, and fragmentation—if you look at those same factors, they are all associated with rapid evolution."

One result of this discovery is that conservation strategies might only preserve species in a very limited sense. Even the term *species conservation,* said Collyer, is problematic. "Evolution always happens. Always," he said. "It's a kind of ignorance, even arrogance to think you can prevent a species from evolving." This thorny tension is particularly evident in the field of conservation genetics, which, as one authoritative text puts it, seeks to "preserve species as dynamic entities." To highlight how difficult this really is in practice, Collyer told me the story of another desert fish case he has focused on in recent years. The Pecos pupfish live along the Pecos River from Texas into New Mexico, but because the river has been dramatically altered by damming and channelization, the fish are now isolated in various habitats. "There used to be constant gene flow and now we've got all these micropopulations with no gene flow and they all have their own evolutionary trajectory," he explained. "How do you fix it? You can augment it but you might also mess up their local adaptions. You could introduce alleles that don't allow them to adapt to their environment. It's hard. It's easier to document the travesty than it is to fix it."

The best conservation policy would be to leave environments untouched, and allow natural processes to take their own course. This is the attitude that has historically guided the conservation movement, which upholds an ideal of "wilderness" as nature untouched by humans. In a perfect world, Collyer said, humans would neither hasten nor retard rates of evolution. Of course, it hardly needs to be said that it is far from a perfect world. A few days after I visited the Lost River, I drove to Santa Fe to visit John Pittenger, the ecologist who has been working on pupfish management since the 1990s. I found the offices for Blue Earth Ecological Consultants, the firm Pittenger cofounded with his wife, on Pacheco Street that runs alongside the Santa Fe Railway line. It was a brilliantly sunny fall day, and we sat in a cheerfully decorated room around a conference table in Navajo-print office chairs. A large geological map of New Mexico hung on the wall. Pittenger explained his opinion that the biggest threat now facing pupfish is climate change. Like most of the western states, New Mexico had been undergoing several years of historically low rainfall. In 2013, with reservoirs nearly empty, state agencies had struggled to keep legally mandated levels of water in the Rio Grande to protect endangered minnows. According to The Nature Conservancy, the Tularosa Basin is one of many high-risk regions in the state when it comes to susceptibility to the stresses of global warming. "The Salt Creek drainage has a much more immediate response to changes in rainfall because the aquifer that feeds it is recharged in the mountain zone," said Pittenger, pointing to the map. "A decade or two of low precipitation can reduce the habitat." I asked his opinion of the genetic distinction between Salt Creek and Malpais Spring. Did he think they represented different subspecies by now? "They've been isolated since the end of the Pleistocene so they are quite a bit different," he said. "It's headed that way though . . . If they last long enough to survive humans."

✳   ✳   ✳

I had begun to wonder why it took so long for scientists to discover that contemporary evolution is both real and commonplace. What about those infamous case studies of natural selection that many of us learn about in biology class, the beaks of Galápagos finches or the peppered moths in England? In the latter instance, scientists have been tracking the evolution of the species for nearly 200 years, ever since the first individuals, generally a light color, were observed in the nineteenth century to be darkening in response to industrial pollution. In the case of the Galápagos finches, the husband and wife team Peter and Rosemary Grant have been observing changes in the beak shape of finches in response to changes in their food supply since 1973. These two cases—let alone all that scientists know from breeding domestic animals like dogs or how quickly bacteria can become resistance to antibiotics—are clear-cut examples of evolution taking place over short time frames. So why the big surprise?

To understand this, I talked with Michael Kinnison, a professor at the University of Maine and an expert in the field of evolutionary biology. Kinnison's scholarly research has focused on the implications of evolutionary rates since the early 1990s, when he was a graduate student studying evolution in chinook salmon populations in New Zealand. In 1999, he coauthored a highly influential paper titled "The Pace of Modern Life: Measuring Rates of Contemporary Evolution." It came at a time when there were a growing number of references to "rapid" evolution in studies but not a lot of context as to what this term really signified. "*Rapid* was all over the place but all it meant was you saw it or detected it in some fashion," explained Kinnison. With his coauthor, Andrew Hendry, now at McGill University, Kinnison brought clarity and consistency to the use of darwins and haldanes, pointing out their mathematical and theoretical shortcomings and strengths, and thereby creating an authoritative reference point for subsequent research. In a bit of foreshadowing, Kinnison and Hendry concluded that article by saying that "the greatest contribution that evolutionary rate estimates will ultimately make is an awareness of our own role in the present microevolution of life and a cautious consideration of whether populations and species can adapt rapidly enough to forestall the macro evolutionary endpoint of extinction."

When I spoke to him, Kinnison explained that the field of evolutionary biology had been distracted from the significance of cases of rapid evolution in modern-day populations of organisms. Examples like the peppered moths were used by scientists to help explain a small part of the biological diversity we see on earth, but in general they were considered rare exceptions of unnaturally powerful human influence, or unique to a particular species. "With evolutionary biology so focused on the species question and the fact that it's difficult to eyeball differences, these other examples, the laboratory studies, the guppies, other critters, were kind of held aside as special cases," said Kinnison. "And there is a self-serving value to those folks who would study those cases in the wild and say they were rare and exceptional examples of evolution in action." When he and other scientists began compiling the documented cases of contemporary evolution, they realized there were many more than anyone had previously thought. "People were not inherently expecting organisms to be evolving in their lifetimes," said Kinnison. "Bacteria? Why not? Pests? There's so many of them, they reproduce annually, sure. But when you start seeing it in salmon and bighorn sheep and trees, and show it's a more general phenomenon, people begin to realize it applies more broadly."

In 2003, Kinnison and Hendry along with Craig Stockwell published a paper called "Contemporary Evolution Meets Conservation Biology," which provided just a smattering of the instances of current evolution observed in the wild. Their list included soapberry bugs, pitcher plant mosquitoes, Pacific salmon, mosquito fish, and sunflowers. They cited a multitude of evolutionary "agents" driving evolution in these species including hybridization, inbreeding, and contaminated soil. Together, they pointed to a new reality for scientists and particularly conservation biologists, who would now need to consider the possibility that their own species of concern were capable of and maybe already undergoing evolutionary changes in front of their eyes, perhaps even as a result of conservation actions. "We challenge conservation biologists to consider evolution in the short term rather than just the long term. This is especially important given that evolution can occur within time frames that are relevant to most conservation plans (decades)."

In the case of the pitcher plant mosquitoes, they cited evolutionary changes to the insects' diapause response as a result of global warming. At the time the article came out in 2003, the evidence for anthropogenic global warming was well established in the scientific literature. And conservation biologists were sounding alarm bells on the impact on species. Now Kinnison, Stockwell, and their group of like-minded researchers were expanding the understanding of global warming's consequences by crystallizing the fact that it can act as a powerful evolutionary agent through its influence on the rate of evolution in populations of animals. Climate change, they argue, has become an unplanned experiment with the biodiversity of the planet. "It is a major selective force and there's a lot of debate out there about how much different species can adapt," said Kinnison. "It's not easy to predict. What might be important for one organism is not the same as another. Seasonality, moisture, it might be for some that climate effects create a competitive balance or mutualism. It's so complicated." Consider, as an example, the ecological relationship between a bird and its prey, in this case a moth, he said. If the peak season of moth hatching and abundance shifts as a result of climate changes, they may no longer be able to support birds that require them as food for their own reproductive cycles. When the speed of evolutionary change is mismatched between any two interdependent species, it could decrease abundance or lead to extinction. "As more species get more and more challenged, society will have to make ethical choices about how far they go [to save them] and what they value and don't value," said Kinnison.

✳   ✳   ✳

The insight into contemporary evolution has opened up an intriguing possibility: that humans could begin to intentionally steer rapid evolution in the direction we want it to go in order to rescue species from endangerment. What if we intentionally introduce selective pressures that lead to stronger, more resilient populations? Could we endow species with characteristics that might allow them to better adapt to climate changes? Kinnison and his colleagues think that this sort of applied evolutionary thinking—or what is also called prescriptive evolution, planned evolution, evolutionary rescue, or directed evolution—could help us understand and respond to extinction threats in the twenty-first century. This is a new, largely unexplored realm in conservation. It requires departing from the traditional "stamp-collecting" mentality, as Kinnison described it, to a process-oriented view. "I don't see many practical ways in which humans are going to extricate themselves from these interactions. If we do, we are often making a choice to let things go extinct."

There are core principles that scientists might use to come up with creative bet-hedging strategies in directing evolution. One such strategy would be actively maintaining or increasing genetic diversity in a population as a way of increasing adaptive diversity. This strategy is particularly useful in habitat restoration. Ecologists generally focus on increasing the population sizes of native plants, but seeding areas with individuals from different plant populations could introduce gene flow and variability to the population. "Maybe you don't want to necessarily get seed from Utah to plant it in Utah. . . . You want to maintain potential in face of global change, so there is genetic variation in there to allow the population to evolve and adapt," explained Kevin Rice, a professor in plant sciences at the University of California, Davis. "If you work in restoration ecology, there is more and more interest in this."

Then there is the possibility that scientists could make more-informed, aggressive bets. For example, endowing organisms with defenses or adaptions to environmental changes, or disease and blights. Researchers are already doing this with American chestnut trees. In the early twentieth century, hundreds of millions of chestnut trees indigenous to the East Coast were nearly wiped out by the fungus *Cryphonectria parasitica,* which produces a lethal chemical for the trees called oxalic acid. Now the SUNY College of Environmental Science and Forestry is growing genetically modified American chestnuts that contain genes from a wheat plant and give them a resistance to the acid. As many as 10,000 of these genetically modified trees could be ready for planting in the species' native range by 2020. This strategy might conceivably work for amphibians. Chytrid fungus, or Bd, has proved fatal to nearly every amphibian species it has had contact with, but if resistant genes could be identified, they might be introduced to at-risk populations to enhance their chances of survival.

In 2007, Scott Carroll, an evolutionary ecologist at the University of California, Davis launched an organization called the Institute for Contemporary Evolution. Carroll's goal is to bring together scientists to focus on the importance of evolutionary forces and the challenges of the Anthropocene, the current period of history in which humans are said to have become a geological force on earth. Ultimately, the institute aims to lay a foundation for a new field called applied evolutionary biology that can bridge research in agriculture, medicine, and conservation biology. The link between medicine and conservation biology is particularly fascinating. The ability for pathogenic bacteria to evolve resistances to antibiotics is a problem that hounds medical science. And the central question researchers are trying to answer is the exact same as it is for conservationists trying to rescue populations undergoing environmental changes and facing extinction, though in one case it's about killing bacteria and in the other case it's about saving animals: How does a population decline or rescue itself through evolution? The Institute for Contemporary Evolution's first conference in 2015 is focused on this intersection between resistance evolution in medicine and rapid evolution in species.

I spoke to Carroll as he was driving home from an entomology conference in Portland, Oregon, with his wife, Jenella Loye. Carroll and Loye have spent years researching soapberry bugs, brightly colored insects found in Asia, Africa, and the Americas that survive by eating the fruits from a genus of tree called *Sapindus*. (Fruit from these trees can be used to make soap, hence the name.) Together they have been tracking the bugs' capacity for rapid evolution and observed populations changing in a matter of decades in response to fruit size, availability, and climate. Since they began their research, soapberry bug evolution, he told me, has become a common element of biology curriculums for undergraduate and graduate students. "What that means is when I was in graduate school twenty-five years ago there was very little recognition that ongoing evolution had anything to do with what's happening on the planet, or people's influence on the planet, or avenues for conservation biology," said Carroll. "Now we've got a whole generation of young biologists who say of course evolution is ongoing."

Convincing conservationists to consider a conscious approach to managing evolution is a particularly difficult task. Wildlife managers, some of whom might work in national parks that explicitly prohibit experimentation with wild populations, are wary to say the least, said Craig Stockwell. "I can tell you this, when I bring it up to managers I work with, they are skeptical. They have been trained to think evolution is off in the future. You can talk them into doing genetic surveys and then you can talk them into moving some fish to restore diversity. They are on board with that. But getting them to think about, well, what if these two populations have different selection regimes, should we move fish between them?"

The underlying problem (in addition to limited budgets and limited time) is that prescriptive evolution stirs a hornet's nest of ethical values in conservation. "The conservation ethic of wilderness being separate from human society and the evolution of wild things with their own value or entities with a right to existence, these are values that are held deeply in conservation, and [directed evolution] starts to become less comfortable for people," said Kinnison. "If we start meddling in the evolutionary trajectory of some organisms and do it in an intended fashion, not like now when it's unintended, we are making a choice about that species, we are manipulating them and choosing their futures." But, he pointed out, "most of what we do in conservation biology is already an evolutionary manipulation."

4

MYSTERIES OF THE WHALE CALLED 1334

Eubalaena glacialis

Thinking about how humans might intentionally direct evolution in species to conserve them reveals how little we still grasp of the complex organic process that has resulted in an animal and the relationship between its genes, behavior, and environment. Add the variable of anthropogenic climate change, and this ecological complexity becomes boggling. Climate change has always been a powerful driver of evolution, but it is difficult to predict the speed of change today and the effects on species. In some cases, these changes might benefit a population, or for those that can't evolve fast enough, it could send them into a tailspin. When it comes to North Atlantic right whales, one of the rarest marine mammals in the ocean, scientists are racing to understand how climate change is impacting these giants, a species that appears to have evolved slowly or not much at all in the last 5 million years. Compounding the difficulty of researching these animals is the fact that most whales have extraordinary life spans. In

2007, biologists found a bowhead whale estimated to be 130 years old with a harpoon point in its skin dating from 1880; some bowheads are believed to live over 200 years. Most whale calves will likely outlive the people studying them. Though it's possible to tag whales with satellite transmitters to track them in the ocean, most of these gadgets fall off or stop working within weeks or months. Scientists trying to get a glimpse of the lives of whales have to be in the same place and same time as the animals in order to glean clues about their biology. In the case of the North Atlantic right whale, just knowing where to look is exceedingly hard.

Prized by whalers for their abundant blubber and gargantuan size, *Eubalaena glacialis* once inhabited the Atlantic from Iceland to Florida and northwest Africa but were hunted with abandon; by the mid-1700s there weren't enough to support commercial hunting. When the industry was outlawed in 1935, the North Atlantic right whale was considered almost extinct, and during World War II, US antisubmarine patrols inadvertently targeted the few animals left along the Eastern Seaboard. But in the 1970s, against all odds, some biologists found a handful of right whales off the coast of New England, a discovery they likened to finding a brontosaurus wandering in their backyard. These whales had a startling lack of genetic diversity. They were one of the most homozygous species on earth, meaning they lacked differences between the alleles on their chromosomes. The right whales also had a ponderous rate of reproduction. Sloths give birth faster and more frequently.

In the decades following their rediscovery, conservationists fought to protect the whales from ship strikes and entanglements in fishing gear along the busy northeastern coast. But after seventy years of international protection the right whales were not bouncing back as quickly as expected. Their evolutionary cousins, the southern right whales, had also undergone heavy commercial exploitation and had low genetic variability, but the population had a reproductive rate three times faster and had rebounded from 1,000 to 6,000 individuals. On top of this mystery, every few years it seems, the North Atlantic right whales do something that baffles scientists and make it difficult to know how best to protect them. This happened most recently in

2013: every summer, hundreds of right whales go to the Bay of Fundy, near Nova Scotia and Maine, to feed, but that year only six right whales showed up. The animals, as long as six-story buildings and weighing seventy tons, just disappeared somewhere in the ocean. When I asked one right whale biologist where he and his colleagues thought the whales went, all he could say is, "We haven't got the foggiest idea."

The study of North Atlantic right whales is comparable to some of the greatest multidecade studies of mountain gorillas, chimpanzees, and elephants. The New England Aquarium's North American Right Whale Catalog contains some 400,000 photographs of whales. Researchers have spent decades conducting population surveys. The animals are one of the only marine species for which there is a genetic profile for nearly every individual, about 80 percent of the population. But after peering into their DNA and spending tens of millions of dollars in research and protection efforts, scientists can still only guess at basic facts of the animals' existence, such as where noncalving whales go to feed in the winter, or where breeding takes place, or where some whales go to feed in the summer. The more I learned about right whales, the more the species began to seem like a potent reminder of something easy to forget: in an era of Google Maps and microchips and unabashed faith in technology's power, some things on earth are so big and complex—oceans, climates, whales—they dwarf our comprehension, let alone our ability to control them, for better or worse.

✳   ✳   ✳

Katie Jackson is a young wildlife researcher whose passion is the survival of right whales. Every winter she captains a twenty-five-foot rigid-hulled inflatable boat with a photographer and an archer to cruise the waters off the Georgia and Florida coastline where mothers come to give birth to their calves. When she finds the whales, Jackson shoots them with a biopsy dart so the skin samples can be used to track the new branches of the species' family tree as they sprout. In twelve years on the job, Jackson has darted around 300 whales.

Right whales are about the same length as humpbacks and grays, but their tubby bodies mean they weigh much more. Even newborn calves are giants, measuring as long as fifteen feet and weighing over 1,100 pounds. In spite of their size, the whales are hard to see on anything but the calmest day at sea. It takes eyes trained to know the difference between a wave and a fluke, or a whale swimming beneath the surface of the water and the shadow cast by a cloud. To increase her chances, Jackson works with two pilots who fly small planes a thousand feet above the sea surface. The aircraft fly in a mowing-the-lawn pattern, heading out thirty-five miles due east before heading back west toward the coast, repeating this path. Jackson splits the difference between the planes and tracks them down the middle of the course, north to south. By the end of the day, the survey team has traversed hundreds of miles.

On February 21, 2013, the weather was mild and Jackson and her team were headed to the Mayport Naval Station to pick up the boat by midday. As they launched into the water, she got a call from Jen Jakush, a lookout in one of the planes. Jakush said she had spotted a female and her calf swimming three miles east off Ponte Vedra, and the pilot was circling the whales so she could try to identify the mother. Jackson headed south toward Jakush's plane when she got another phone call. "It's #1334," Jakush said.

Jackson was surprised. "And then I was excited," she said, "and scared."

The whale known as 1334 (an identifying number given to right whales by the New England Aquarium) had confounded biologists for thirty years. She was first seen off the southern coast back in the early 1980s and reappeared there periodically. But unlike the others, 1334 did not show up in the Bay of Fundy in the summers with the rest of the right whales. No one saw her again, until she appeared in Florida three years later with a new calf. And then the same thing happened three years later. For the next thirty years, 1334 was the most prolific right whale mother known to researchers, giving birth to nine calves like clockwork. 1334 gave birth during years when biologists saw calving rates stall and even decline in the general population. In 2000, she was the *only* right whale to give birth to a calf, off of

Ossabaw Island, Georgia. But where she fed, mated, or migrated in between giving birth, no one knew.

This mystery had deepened in 1989, when 1334 was spotted from a boat in the Labrador Basin, a depression in the northwestern Atlantic floor where the icy water runs two miles deep. There, some 500 miles southeast of Greenland and 650 miles east of Labrador, south of an old historical whaling area called Cape Farewell, 1334 was swimming with an eight-month-old calf. Their 2,660-mile journey from the calving grounds to the Labrador Basin is among the longest documented migrations of any right whale. 1334 was not the only female who went missing each summer, but she became the most famous. (Another female whale, known as Rat, had been sighted over 200 times but appeared in the Bay of Fundy for only two months in 1997.) "A handful of whales would show up here to nurse calves and no one knew them," said Jackson. "We never saw them in other places and it brought about this whole new question of, *where* are these whales? And are there more whales there? 1334 became the epitome of that idea." In all, about a third of the female right whale population never brought their calves to the Bay of Fundy. Biologists started calling the missing animals "non-Fundy" whales, a name that belied the vacuum of information around them. Gathering genetic material from their calves was one way to begin filling the vacuum; if they could figure out the calves' paternity, perhaps it would reveal an important clue.

Jackson wasn't surprised 1334 had shown up with a calf, but she was surprised that the whale was only three miles offshore. 1334 had never been seen so close to land and no one had ever been able to get a genetic sample from her. The closest anyone came was Jackson herself in 2009 when she was seven years into surveying whales. Her team had approached 1334 some thirty-five miles offshore, but when they tried to get close for the archer to take a shot, she dove underwater with her calf for ten minutes and then reappeared a hundred yards away. The whale did the same thing on their second approach and it seemed like no matter where the team predicted she would resurface, 1334 was somewhere else, out of reach of their crossbow. Jackson sensed that the whale was inserting confusion into the chase, moving quickly and unpredictably to avoid the boat.

For hundreds of years, right whales had been sought and killed by whalers precisely because they didn't display this kind of behavior. Right whales were considered docile and easy to kill: they moved slowly and submerged for brief periods before resurfacing nearby. When the *Mayflower* anchored in Provincetown Harbor, one passenger, likely the Englishman William Bradford, wrote about the abundance of right whales and their apparent lack of fear of the boat's presence: "And every day we saw whales playing hard by us, of which in that place, if we had instruments and means to take them, we might have made a very rich return; which to our great grief, we wanted." Jackson and her team gave up on 1334 and she disappeared only to reappear now.

The National Oceanic and Atmospheric Administration enforces strict protocols for approaching endangered whales. In order for Jackson to take a sample, she had to make a second identification to confirm it was really 1334. Usually this entails using a zoom lens to get a photograph of the whale's head and then retreating to compare the photograph with a catalog of whales on the boat. Jackson didn't want to risk losing her opportunity to get a biopsy. "I knew if I got a good enough look at her, I would know," said Jackson. "There are some whales that are infamous, they have faces that you can't forget." Luckily, as Jackson steered toward the right side of the whale, 1334 lifted her head and rolled, giving Jackson a glimpse of some scarring and her callosity pattern, a buildup of tissue on top of a right whale's head. This tissue is host to white, crab-like parasites, known as "whale lice," that contrast with the whale's black skin, creating a distinct pattern. 1334 had one large teardrop-shaped callosity with two small circles underneath. "I'm sure," Jackson said to the archer when she saw it. "Take whatever shot you can get."

Tom Pitchford aimed his crossbow, his finger on the trigger. The bow had a 150-pound draw weight and his arrowhead was a one-inch, stainless steel hollow cylinder, sharply beveled at the end in order to penetrate the animal's skin. Inside the cylinder were three backward-facing prongs to hold the piece of skin and blubber in place, and a foam core to prevent the tip from penetrating too deeply. The fletching of the arrow was a piece of foam, so that the whole contraption would float in the water after it rebounded off the whale. Pitchford aimed at the right side of 1334 and pulled the trigger, sending the arrow into the whale's epidermis. Jackson steered the boat to grab the arrow from the water, and then focused her attention on the calf. But now the mother was more evasive. Jackson made a couple of fruitless approaches as both animals submerged. Finally, Pitchford got close enough and took a shot at the calf, landing the arrow in the calf's left side before it fell into the water, where the survey team grabbed it. To Jackson, their luck felt almost unbelievable. "It was a moment where everything came together and worked in our favor and we don't have many of those," she said. "That was the one and only sighting of 1334 all season."

Pitchford put the biopsy samples into a cooler and then processed them back at their field station, putting gloves on to prevent contamination with his own DNA and cutting the pieces of black epidermis with a scalpel into even smaller pieces. Each one went into separate vials of preservative, and a couple of weeks later, one of them was delivered to Brad White, a molecular biologist at Trent University's Natural Resources DNA Profiling and Forensic Centre. White had been analyzing right whale DNA for thirty years. He knew that if you compare two right whales' DNA, they practically look like twins. But he also knew about 1334's strange wanderings and remarkable fecundity. Maybe her DNA had something different, something that would help them solve the riddle of the species' survival.

$$* \quad * \quad *$$

For 400 years, the genetic clue to unlocking the North Atlantic right whale's evolutionary history lay buried beneath a shipwreck in the subpolar sea off the coast of Labrador. Then in 1978 Selma Barkham pointed archeologists to the site and told them to look under the water. A widow with no university degree, the mother of four children, Barkham was an unlikely candidate to make one of the great discoveries in maritime history and, in the process, prove one of the earliest links between the Old and New Worlds.

Born in London in 1927, for many years Barkham had scraped by financially, teaching English in Mexico and Spain. But her maiden name, Huxley, revealed an illustrious intellectual and scientific ancestry, and her character was in keeping with the family's tradition of producing impressive autodidacts. Her maternal grandfather was a prime minister in Quebec, and her father was Michael Huxley, founder and editor of *Geographical Magazine*, the publication of England's Royal Geographical Society. Her father's cousins included Aldous Huxley, author of *Brave New World* and *The Doors to Perception*, and Sir Julian Huxley, a biologist and the first director general of UNESCO. Aldous and Julian's grandfather was Thomas Henry Huxley, the eminent scientist whose formal schooling had ended at the age of ten but who received many of his field's most illustrious medals before his death in 1895. As a teenager, Thomas Huxley taught himself German, Latin and Greek, and became an expert in anatomy, invertebrate and vertebrate zoology, paleontology, theology, and medicine. He became famous for his fierce defense of the theory of evolution, earning himself the moniker "Darwin's bulldog."

His great-niece Selma Barkham would receive many accolades later in life too. In 1981 she was awarded the Gold Medal of the Royal Canadian Geographical Society and was appointed a Member of the Order of Canada. But her career couldn't have begun more modestly. She was first a schoolteacher and then a librarian at the Arctic Institute of North America at McGill University in Montreal. In 1953 she married a British architect, Brian Barkham, a man with a passion for the Basque country where he had studied Spanish architecture. After they were married, Brian brought his wife to his favorite country and it was there, in 1956, that a priest mentioned a long-forgotten connection between the Basques and Canada. He told them that in the sixteenth century, the Basques had sailed across the Atlantic in search of fish along the Canadian coastline. This conversation and piece of trivia lodged in Barkham's mind. She gave birth to four children in Ottawa, but then her husband passed in 1964. Her son Michael was ten years old when his mother announced she had a plan for the family's future: they would leave Canada and move to Mexico. She would teach English and, in turn, learn Spanish. Then they would all go to Spain, where she would follow through on an idea she had harbored for years: to research and write about the Basques in Canada. "The extraordinary thing is, there she was, a widow with four children, no family fortune behind her, but she was the type of person who decided she would do this research," said Michael. "I think it allowed her a chance to combine her intellectual interest with this personal element of her husband, who she had been to Basque country with and who still had friends there."

After living in Mexico for three years, the Barkham family took a cargo ship to Bilbao during the waning years of Franco's dictatorship. But when Barkham arrived, she found out that her request to the Canadian government for research funding was denied because they thought the project wouldn't reveal anything of new historical value. To support her family, Barkham once again taught English and conducted her research on the side. Over the next decade she spent hundreds of hours in libraries and archives in the cities of Seville, Lisbon, Madrid, and Tolosa. It was true in a sense that historians *had* already established that the Basques pursued fishing across the Atlantic, but what Barkham found in previously unexplored contracts, wills, and insurance policies was a far richer maritime economy dating back to the early 1500s and focused not on cod, but on whales.

The Basques were expert seamen and had hunted whales going back to the twelfth century along the Cantabrian coast and as far as the Irish Sea. Shipbuilders and entrepreneurs, they were soon going even farther abroad, to Newfoundland, for hake and cod. More than likely it was the French Basques who were the first to return in the early 1500s with reports of whales in the land they called *Terranova*. Barkham discovered records showing dozens of ships that made the journey, landing in the Strait of Belle Isle, an icy, foggy body of water that separates the Labrador Peninsula from the island of Newfoundland. They came for *la venida de las ballenas*, the "coming of the whales" in June and July, and established camps near the deepwater harbors along the strait where their large, 450-ton galleons could be anchored. The hunting season soon grew to include *el retorno de las ballenas* in September and October.

The Basques brought harpoons and red tiles to build roofs over ovens and furnaces, heavy copper cauldrons to render whale blubber, and metal hoops for coopering the barrels that stored the oil on the return voyage. They used a particularly cruel but effective method to hunt whales. First they harpooned young calves to mortally wound them, knowing the mothers would stay close to their injured young, and the harpooner could then kill her. A single ship might return to Spain with as many as a thousand barrels of oil rendered from a dozen whales that could be sold at a high profit. By today's standard, the owner of a whaling galleon stood to become a millionaire after just two or three whaling expeditions, and then he could sell his ship for even more profit. Between the 1540s and 1620s, the Basques were harvesting on average three hundred whales per year and bringing around 15,000 barrels back home. Right whale oil burned across Europe.

In 1974 Barkham went to the small Basque mountain town of Oñati and found an archive there that had been virtually untouched for centuries. As she looked through the records, she found a peculiar document, a lawsuit regarding a whaling ship called the *San Juan*. In 1565, the 204-ton galleon had been loaded with cargo and ready to sail back to Spain when an unexpected storm broke the ship's moorings and drove it aground where it sank. Intrigued, Barkham scoured contemporary maps of Labrador and compared them to the old maps she found in the archives. At the site of what the Basques called *Les Buttes* in the sixteenth century, on the north side of the Strait of Belle Isle, was a little harbor town called Red Bay, named for the red granite bluffs nearby. This was where, she concluded, the *San Juan* had sunk. Barkham applied for a grant from the Royal Canadian Geographical Society, and in 1977 she launched an archeological survey to explore the harbors along the strait for evidence of Basque whalers.

Michael Barkham was eighteen years old and went with his mother on the trip. In the spring and early summer, icebergs still float through the Strait of Belle Isle and the coastline gets violent weather—gales and freezing rain. "It was an adventure getting to Labrador," he said. It didn't take long before they found what they were looking for. "Sure enough, walking along the coast and surveying the harbors, we found piles of whale bone. A lot of locals thought they were a hundred years old, they had no idea they were 450 years old. And there was Iberian roofing tile just lying on the beach." Residents of Red Bay thought the tiles were red bricks from England and told Michael they ground it up and painted children's faces with it. They weren't the only ones who had misidentified the tiles that littered the coastline. Sir Joseph Banks, the English naturalist who had traveled with Captain James Cook on the first voyage of the *HMS Endeavor,* believed the Vikings left the tiles there. James Tuck, an important archeologist of Newfoundland and Labrador's prehistoric history, had spent years along the coast excavating 10,000-year-old Native American sites, but he never thought the red tiles were significant.

With so much evidence to support her research, Barkham was more confident than ever about the *San Juan*. She told an underwater archeologist with the Parks Canada Agency, Robert Grenier, that she was positive the *San Juan* was real, and gave him an estimation of where the wreck was located—on the north edge of Saddle Island, a small landmass in Red Bay's harbor. In 1978, Grenier took a small marine unit to Saddle Island and sure enough, just thirty feet beneath the surface of the water, they found ship timbers protruding through the silt.

The excavation of the *San Juan* took five years. Divers descended during the warmer months, before cold temperatures and encroaching ice made the work impossible. The ship had been immaculately preserved in its watery grave. By the time they finished in 1985, the unit had spent 140,000 hours underwater dismantling and raising 3,000 timbers. At the surface, archeologists diagrammed each piece of wood and documented its shape, tool marks, and wear. A 1:10 scale model of the ship was assembled and when the team was finished, each piece of timber was brought back underwater and buried at the original site to maintain its preservation. Among the shoes, wooden bowls, harpoon shafts, and barrels that divers discovered were twenty-one whale bones preserved by the sea. The bones were humeri from whale flippers, and in 1986 an osteologist identified half of them as belonging to bowhead whales and the other half as belonging to right whales. Barkham knew the bones were a significant discovery for biologists, giving them a rare snapshot of the species' history that had been lost to the oceans and time.

But she could not have predicted that they would end up radically changing scientists' understanding of the species.

Since their discovery of a significant population in the late 1970s, biologists had believed that right whales had been brought to the brink of extinction by colonial whaling, an industry that exploded during the seventeenth century and resulted in the harvest of at least 5,500 right whales. Based on this number, biologists estimated there had to have been at least a population of 1,000 whales in the 1690s, and by the time hunting was banned, as few as seventy left. Following the discovery of the Basque whaling enterprise, biologists could date the right whale's exploitation and population bottleneck hundreds of years earlier than they previously thought. Scholars working in Barkham's footsteps estimated that the Basques killed 25,000 to 40,000 bowhead and right whales. Based on this estimate and the whale humeri found underneath the *San Juan*, marine biologists calculated the population size of the North Atlantic right whale at around 12,000 to 15,000 individuals before the Basques arrived. In 1991, the National Marine Fisheries Service presented a conservation plan for the North Atlantic right whale and used this estimate as a measuring stick for the recovery of the species. The whales had a long way to go; in the 1990s their population was estimated to be a few hundred individuals at most, and they were projected to disappear within two centuries if nothing changed.

In the early 2000s, Brenna McLeod, a molecular biologist, was searching for a graduate research topic at Trent University. McLeod's interests were split between marine mammals and anthropology, and her adviser, Brad White, pointed her to the old bones discovered beneath the *San Juan*. Preliminary analyses had been undertaken by another researcher, Toolika Rastogi. The bones were so well preserved, it was easy to extract high-quality genetic material, but surprisingly, Rastogi's analysis showed only one bone from a right whale, and the rest from bowheads. The lab needed more work done, and a larger sample set, White told her.

McLeod threw herself into the project. She decided to genetically analyze the right whale bone at a number of microsatellite loci, the molecular markers in DNA scientists use to understand kinship and population. She chose loci that had already been analyzed in the modern-day population, figuring that if the North Atlantic right whale had undergone a population bottleneck as a result of Basque hunting, as biologists believed, then she would find rare or lost alleles in the bone from 1565. Instead, McLeod got a second surprise. The bone might as well have been from a living right whale swimming off the coast of Florida. "The genotype looked identical to animals in the population today," said McLeod. "You'd expect that if a lot of variability had been lost, it would look much different. But individual whales in the 1500s were very similar to today. They didn't go through a huge bottleneck from whaling as previously believed." These implications were startling. At some point in their evolutionary history, maybe as a result of climate change, North Atlantic right whales might have been reduced to as few as eighty-five individuals, then rebounded, only to be reduced again by whaling. Whatever caused their extreme lack of genetic diversity probably didn't originate with the Basques *or* later whaling fleets.

McLeod started presenting her initial findings at conferences, and the response she got was anger. "People believed that the population was once large and that humans did this to them," she said. Among the criticisms of her work was that she hadn't used a big enough sample size, and this had skewed her findings. So McLeod decided to keep going, turning her research into a PhD and spending three subsequent summers hunting whale bones along the Strait of Belle Isle. At first she surveyed the few beaches that were accessible from the only access road along the coast. In 2004, the biologist Michael Moore of the Woods Hole Oceanographic Institute offered his boat, *Rosita*, to help her get to more beaches. Moore had spent a sabbatical year on the *Rosita* looking for right whales in old historical whaling grounds (he didn't find any), and he was intrigued by the possibility of finding more bones that could help solve the puzzle of the species' DNA.

With the advice of Selma Barkham, now living in England, as well as some old maps, McLeod and Moore studied 150 miles of coastline and identified dozens of Basque whaling sites. Along with Moe Brown and Yan Guilbault of the New England Aquarium, they sailed for five days, from Massachusetts to Red Bay where the *Rosita* arrived under thick fog. For the next ten days, Moore anchored the boat in the different harbors and McLeod, armed with a portable drill, searched under moss and rocks for bones. Brown, a former physical-education teacher in Montreal who became a crusader for right whale conservation, scuba dived along the shore to look for bones in the water. By the end of the trip, they had 200 bone samples.

In 2009, McLeod and her cohorts published their findings through the Arctic Institute of North America. Of 364 bones, there were one blue whale, one fin whale, two humpback whales, and 203 bowhead whales. None of the samples belonged to a right whale. Rather than solving the puzzle, the results were confounding. Bowheads are a species thought to be limited to the icy waters of the eastern Arctic. What were they doing so far south in the sixteenth century? And if there were no right whales around for the Basques to kill, why did the species have so little genetic diversity for over 400 years? "Right whales clearly had something else going on," said McLeod. "Some sort of climate variability or environmental impact. We just don't know. I don't think there were always 300 whales in the population. But there were never 40,000."

One theory that McLeod and her fellow authors toyed with was that the "Little Ice Age"—a period from 1300–1850 CE when sea surface temperatures dropped, glaciers advanced, and the Arctic front expanded—shifted the migration patterns of bowhead and right whales. Maybe bowheads pushed farther south with the cold front, while right whales were forced from their northerly feeding grounds into warmer waters. But this didn't explain genetic drift—the random loss of variation over generations—seen in right whales. It would have required many Little Ice Ages, tens of thousands of years of oscillating climates, each one impacting the whales' survival and contracting the population, to create the genetic drift and lack of diversity seen in right whales. Underlying all these questions was how the right whales had survived for so long with so little evolutionary potential.

Right whales, it turns out, aren't the only instance of species surviving thousands of years with extremely low genetic diversity. The social tuco-tuco, for instance, is a rodent endemic in the grasslands of Argentina that has survived over 950 years with a single mitochondrial haplotype, the set of DNA variations inherited from mothers, and still manages to produce viable offspring. The Madagascar fish eagle survived with around 120 breeding pairs for the past 2,800 years. In 2007, French and Canadian biologists presented their research on Amsterdam and wandering albatrosses, two species that diverged from a single ancestor about a million years ago and have survived for hundreds of thousands of years with startlingly low genetic variation. But remarkably, the two species have high reproductive success rates, and in the case of the Amsterdam albatross, it rebounded from a severe population bottleneck of just five breeding pairs in the early 1980s. "Albatrosses," the authors of the study wrote, "appear to challenge the classical view about the negative impacts of genetic depletion on populations."

What explains this mystery? It turns out that right whales and albatrosses may share a similar behavior that helps their resilience: both species, it appears, have the ability to avoid inbreeding. In the case of the whales, this could have something to do with their complicated courtship rituals. Female right whales are incredibly promiscuous and throughout the year, a female will roll on her back and call to any nearby males, who swim from miles away to congregate around her. Each male attempts to get close to the female, jostling others in the process, slapping the surface with their flippers and tails, and blowing water. When the female turns right side up for air, the males swim underneath her and try to mate. This maelstrom, called a surface active group, can last for hours but it's not aggressive. Male right whales don't fight each other for the right to mate; instead they practice some of the strongest sperm competition known in the animal kingdom. When it comes time to mate with a female, numerous males might inseminate her, and their sperm race to fertilize her egg. Right whale biologists know by looking at genetic samples gathered by researchers such as Katie Jackson that this fertilization is only successful when the male has dissimilar DNA from the female. If inbreeding occurs, there is a high rate of spontaneous abortion. The result is that right whale offspring have higher levels of genetic variability than scientists would expect from a population with such a minuscule gene pool. Amazingly, each time a male and female right whale successfully produce a calf, they increase what's called the effective population size of the species, the number of individuals who can contribute genes to the next generation and slow genetic drift.

Archeologists' discovery of the whalebone beneath the *San Juan* and McLeod's analysis upended the understanding of right whales and how to save them. Five hundred years of whaling had dramatically reduced their abundance, but the animals' mating behaviors likely helped them survive this bottleneck. Indeed, they may have always had a relatively small population. This was both bad and good news. The good news was that if right whales had a small population for thousands of years, the modern-day whales were probably more stable than they appeared. But the bad news was that scientists' expectations of how fast the population could rebound from their most recent brush with extinction were dashed. The North Atlantic right whales were and remained extraordinarily rare, and this demographic reality coupled with their paltry genetic variance would be a major factor affecting their ability to evolve and adapt to whatever changes to their environment awaited.

* * *

In 1996 the International Whaling Commission held a symposium in Oahu, Hawaii. The topic was a novel but pressing one for the commission—the potential impacts of climate change on the world's cetaceans. There had been recent reports from the Antarctic that ice shelves were crumbling and climate scientists were warning that the coming century could bring record-high temperatures. Sitting in the audience was Bob Kenney, a research scientist at the Graduate School of Oceanography at the University of Rhode Island.

Kenney is part of a core group of New England biologists who have studied North Atlantic right whales since their rediscovery in the 1970s. In conservation, it's not unusual for personalities to clash and competition for funding to be ferocious. In this respect, the North Atlantic right whale conservation effort has been unique. In 1986, researchers from five institutions banded together to share their findings and apply for funding together as the Right Whale Consortium, and later they expanded to include more than 100 institutions, adding government agencies, environmental lawyers, and even members of the shipping and fishing industries. Today the consortium remains a remarkably fraternal group, producing a public quarterly newsletter detailing the latest events, research, and publications related to right whales that might include some poetry or a moving testimonial detailing a particularly special whale sighting. Every year, the group meets at the New Bedford Whaling Museum, an event that is considered a must for anyone who does anything with right whales.

Kenney, a genial man with white whiskers and hair to match, is coincidentally a descendent of the same William Bradford who described the presence of whales around the *Mayflower*. The way he tells the story, he got involved with right whales pretty much by accident. In 1978 he arrived at the University of Rhode Island as a graduate student to meet his thesis adviser, Howard Winn, a pioneer in researching humpback whale songs and marine acoustics. Kenney walked into a room looking for Winn and instead found a group of people talking about what would become a landmark study of marine mammals in the North Atlantic, the Cetacean and Turtle Assessment Program. Winn, it turned out, was the scientific director of the effort, and he brought Kenney, whose interest was piqued, on board.

For the next three years Winn's team surveyed animals up and down the northeastern seaboard, and in May 1979 they conducted what they called a "minimum right whale count," in which five boats and six airplanes were deputized to look for right whales over a three-day period from eastern Long Island to Nova Scotia. Kenney rode in a lobster boat from Nantucket, on the eastern side of what is called the Great South Channel, which turned out to be the only place right whales were spotted that spring. What the scientists didn't know was that those whales probably represented a significant percentage of the entire population, estimated to be a couple hundred individuals at most, that was spring feeding in Cape Cod before making their way north toward Canada.

Initially, no one was sure where the whales went once they left the Bay of Fundy. Then in 1983 two Canadian researchers, Randy Reeves and Edward

Mitchell, were sleuthing in old whaling logbooks from the 1800s and found reference to a New Bedford whaling schooner that killed a right whale near Brunswick, Georgia. Based on this clue and a few others, a group of pilots from Delta Air Lines volunteered to survey the southeastern coast in their private planes to search for right whales and quickly found around a dozen mothers in the temperate waters with their newborn calves. Still, no one knew whether these were the same whales that appeared up north. The biologists started photographing each animal sighting and recording their distinct callosity patterns, and the New England Aquarium organized these records to create the Right Whale Catalog. As the catalog grew, so did biologists' ability to distinguish the whales from one another. Sure enough, many of the Bay of Fundy whales were appearing 1,500 miles south each winter, including the pregnant females who gave birth and nursed their young in the warm water. Today, the Right Whale Catalog contains photographs of over 600 individuals dating back to 1935.

In 1996 there was no direct evidence that climate change was negatively affecting right whales in the North Atlantic. But the idea that future climatic shifts might impact the population was alarming. The animals were already prone to ship strikes; it was not unusual to identify them by the scars of propeller blades, and each year a couple of shredded carcasses would wash ashore. Even more alarming was that the intervals between females giving birth seemed to be extending from three or four years to five or six, and the survival rate of adult females appeared to be slipping. What was going on?

In Oahu, Kenney attended a panel discussion given by scientists working in Antarctica. They had been witnessing short-term changes in climate that seemed to be affecting krill abundance and, in turn, the reproduction rates of seals and penguins. The scientists presented a graph showing variability in the Southern Oscillation Index, the measurement of the air pressure difference between Tahiti and Darwin, Australia. The graph jerkily bounced up and down and it looked strangely familiar to Kenney. "Every time someone put the graph up showing the numbers, it looked to me like the same graph that I would draw if I plotted how many calves the right whales were giving birth to," said Kenney. "Which made no sense at all."

Kenney's curiosity was stirred enough that when he returned to Rhode Island he tracked down the data on the Southern Oscillation Index and did a statistical test against right whale calving rates. What he found was that there was a significant correlation between the two.

Over the next ten years, Kenney began investigating the relationship between climate phenomena and calving rates, next looking at the North Atlantic Oscillation, or NAO (the difference in atmospheric pressure between Iceland and the Azores), and then the Gulf Stream Index (the latitudinal position of the stream) dating back to the 1980s. All three atmospheric cycles could be correlated to calving rates if he allowed for a one- to two-year time lag. But beyond that, it was hard to know what to make of the data. What was it revealing?

Then Kenney got an e-mail from Charles Greene, director of the Ocean Resources and Ecosystems Program at Cornell University. Greene and his graduate student, Andrew Pershing, had been researching the response of a tiny crustacean called *Calanus finmarchicus* to fluctuations in the North Atlantic Oscillation. What Greene knew, and wanted to discuss with Kenney, was the fact that *C. finmarchicus* is the principal food source for North Atlantic right whales.

C. finmarchicus belongs to a subclass known as copepods, meaning "oar feet" in Greek, and the organism is among the most abundant form of zooplankton in the North Atlantic. The creatures look like lobsters in miniature, with ten legs and two long antennae that they use to swim like humans would use their arms to do the breaststroke. Adult *C. finmarchicus* measure just three millimeters long, so small in comparison to a whale that researchers have likened the copepod-whale relationship to humans relying on bacteria for sustenance. Right whales prefer *C. finmarchicus* to any other food because in their juvenile stage the crustaceans develop an oil sac filled with wax esters, a mix of fatty acids and fatty alcohol rich in energy. Using a method called ram filter feeding, a right whale swims forward and seawater enters through an opening in its mouth called the subrostral gap, passing over the tongue and flowing through the space between the whale's lips and its baleen plates. This sulcus narrows in size from front to back, causing the seawater to accelerate in speed as it passes to exit through gaps in front of the whale's eyes. The difference between speeds in the sulcus and the interior of the mouth has a hydrofoil effect, creating pressure that actually pulls water through the baleen, the whale's comb-like "teeth," a marvelous adaptation that increases the whales feeding efficiency. The oil sacs of *C. finmarchicus* are reddish-orange in color, and when they aggregate at the surface of the ocean, they stain the water. Herman Melville described the sight of right whales feeding on the crustaceans as "morning mowers, who side by side slowly and seethingly advance their scythes through the long wet grass of marshy meads; even so these monsters swam, making a strange, grassy cutting sound; and leaving behind them endless swaths of blue upon the yellow sea."

But even the amazing hydraulics of the right whale's mouth can't justify the enormous amounts of energy it requires for a seventy-ton animal to feed on crustaceans the size of rice. Kenney spent years of his professional career researching what makes certain places like the Bay of Fundy the preferred feeding grounds for right whales. *C. finmarchicus* is abundant there, but it is one of the most abundant organisms throughout the entire North Atlantic. What is special about the Bay of Fundy is that the copepods are organized into highly concentrated patches by ocean currents. These patches are like pancakes, each layer of zooplankton hundreds of meters wide but only a few meters thick at most, one on top of the other. The aggregations, Kenney believes, have more to do with physics than biology. "These copepods are packed into tight little patches and the whales need to find patches where it pays to spend the energy to open their mouth and filter them out," he explained. "It's not so much how many *Calanus* there are, it's how strong the currents are, the stratifications in the water and wind."

Kenney and other researchers once estimated how many calories a right whale needs to survive. The number is staggering. At minimum, a right whale has to consume 400,000 calories per day. For a pregnant female, however, this number could be as high as 4 million calories, or roughly 2.6

billion *C. finmarchicus*. Reproduction is an enormously expensive event, Kenney likes to say. Even more expensive is nursing a calf for twelve months until it is big enough to wean and feed itself. During that time a calf doubles its size to around two dozen feet and weighs as much as an armored vehicle. It probably costs three times as much for a mother to feed the calf as it does to grow it inside of her. The result is that if a female right whale has not been able to build her blubber reserves, she will delay pregnancy until her stores of energy are large enough to support giving birth.

Kenney, Greene, and Pershing began assembling their data, and what they found is that when the NAO is positive, meaning that the difference between atmospheric pressure in Iceland and the Azores is strengthened, *C. finmarchicus* is abundant and the rate of calving for a female is every three to four years. But that calving rate lengthens when copepod aggregations decrease. The same year that Kenney was in Oahu at the International Whaling Commission symposium, for instance, the NAO underwent its largest single-year drop in the twentieth century. Two years later, the typical lag time between oceanographic responses to changes in the NAO, *C. finmarchicus* declined in the Gulf of Maine, and the result was a plummeting calving rate among right whales. In 2003, Kenney, Greene, and Pershing, along with Jack Jossi of the National Marine Fisheries Service, published a paper establishing the linkage between the oceanographic forces, climate variability, and the reproduction of North Atlantic right whales. The direct impact of climate change on whales came from prey availability and whether the atmosphere produced conditions in which *C. finmarchicus* could enter into the Gulf of Maine and aggregate into dense patches.

"The climate-driven changes in ocean circulation observed over the past forty years have had a profound impact on the plankton ecology in the [Gulf of Maine]," they wrote.

During the centuries of commercial whaling, humans were clearly the major source of right whale mortality. Since the cessation of whaling, most attention has remained on human activities, such as shipping and fishing, which directly affect mortality rates. Because right whale population recovery is more sensitive to mortality rates than birth rates, conservation efforts to reduce collisions with ships and entanglement with fishing gear are appropriate. In addition, however, we suggest that attention also should be focused on the effects of climate variability on right whale calving rates. Failure to account for these effects of climate may cause us to underestimate the conservation efforts required to ensure recovery of the North Atlantic right whale population.

Kenney and his coauthors believe that long periods of negative NAO will depress calving rates to levels that could not sustain the right whale population with present levels of human-caused mortality. There is also the possibility that a positive NAO over time could benefit right whales by increasing food sources. But climate variability—the dramatic swings between negative and positive NAO from year to year—could easily throw the struggling population into a reproductive tailspin. "Ultimately, our ability to assess the long-term prospects for North Atlantic right whale recovery may only be as good as our ability to predict regional climate variability and change in the Northwest Atlantic," wrote the researchers.

The larger role that baleen whale species, of which right whales are one, play in the ecosystem and regulating climate has only recently been revealed to biologists. Although few in number relative to other marine species, whales likely act as pumps for recycling nutrition around the ocean, feeding on copepods and krill and then releasing fecal plumes and urine that fertilize water for phytoplankton to grow, which then supports the ocean's food chain from the bottom up. Whaling disrupted this system by reducing whale populations, in turn likely influencing carbon storage and fish stocks. And so just as climate change may be negatively affecting North Atlantic right whale populations, their recovery may be needed more than ever to help stabilize oceans against the potential impacts of climate change.

These problems pose a nearly impossible challenge for right whale conservationists. The anthropogenic forces driving shifts in the atmosphere and oceans are beyond the powers of any one organization or alliance to control. The options available to other conservation efforts like translocation and captive breeding are impractical when it comes to right whales, whose relationship to their environment is so specifically aligned. There is no aquarium where seventy-ton right whales can be bred in captivity, and so far there is no way to provide right whales with an alternative food source.

There is the possibility that if the species' genetic diversity could be increased, it would help increase their population size and rate of reproduction. Some scientists have toyed with the idea of introducing a South Atlantic right whale into the northern population, but the logistics of the effort are overwhelming. The southern whales' blubber effectively prevents them from making such a journey into warmer waters around the equator. There was one instance, 1 to 2 million years ago, when at least one southern right whale migrated north, a fact that researchers at the University of Utah surmised by studying the genes of whale lice. Clearly, conservationists can't rely on a once-in-a-million-year event.

But there is another option: scientists could intervene much more brutally into the whale population's gene pool. As more data is collected on new calves' DNA and their paternity, scientists have discovered a portion of male right whales that aren't fathering offspring. Given the mating system is one of sperm competition, this could mean that some males are dominating the surface active groups but contributing unviable sperm that lead to fewer pregnancies for the females. "The question would be," explained Brad White at Trent University, "if you identify those dominant, problem males, would we take the action of killing them?" It's a sobering possibility that the best way to ensure the whales' survival in the long term might turn out to be culling individuals from the small population now.

* * *

In the spring of 2014, the laboratory at Trent University in Ontario began building 1334's genetic profile from her biopsy sample. Brad White had a hunch about 1334. Was it possible she possessed a genotype that allowed her to give birth to calves, irrespective of good or poor nutrition? Of all the female right whales, 1334 was the most fertile and consistent, producing nine calves irrespective of North Atlantic, Gulf Stream, or Southern Oscillation indexes. Initially, the analysis of 1334's nuclear DNA showed extreme homozygosity in her alleles, meaning there was a lack of strong genetic difference between her mother and father, typical among the right whale population. "In a more variable species you wouldn't see this as much," noted White, as he looked through the graphs of 1334's DNA, nearly every one showing identical alleles at one location on her chromosomes after another.

White's hunch about 1334's fertility and genes is the result of a growing body of research around a very different species, but one that whales share a common ancestor with 50 million years ago: the milk cow. In the last forty years, the average milk production for a single dairy cow has doubled because of genetic selection for this trait. But it turns out that a dairy cow's fertility declines with the more milk she produces. There is a strong economic motivation to unlock the genetic relationships between dairy cow milk production and fertility. In 2009, researchers led by the National Institutes of Health and the Department of Agriculture announced they had successfully outlined the genome of a Hereford cow from Montana, an effort that took 300 scientists six years. The cow, known as "L1 Dominette 01449," turned out to have around 22,000 genes, 14,000 of them common to all mammals. Subsequent research identified genes and chromosomal regions in the dairy cattle genome associated with traits like milk, fat, and protein yields, and, most significant in White's opinion, reproduction. "Based on the results in cattle, we know there are certainly genes that make animals more sensitive to lower nutrition and becoming pregnant or maintaining a pregnancy," he said. White believes that by sequencing the North Atlantic right whale genome and looking in those areas that have been shown to affect milk cow reproduction, they might begin to understand how genetics is related to nutrition, environment, and reproduction in the whales. In 2013, he sent genetic material for the North Atlantic right whale to molecular biologists at the University of Utah, who had already sequenced the South Atlantic right whale genome, and in 2014 they finished a draft sequence and began preliminary comparisons to both dairy cattle and southern right whales.

The Trent University laboratory was also interested in 1334's mitochondrial DNA, the material that she inherited from her mother. Mitochondrial DNA is transmitted through the egg cytoplasm, the substance between the nucleus and membrane of cells. Some parts of this genetic material evolve as much as ten times faster than nuclear DNA, and it can reveal a lot about the behavior of animals. Like migratory birds, right whales and other baleen whale species of the *Mysticeti* suborder have strong fidelities to the places they take their young calves to feed. A mother uses the same "nursery" that her mother used, and she passes that fidelity to her daughters, creating a lineage through the generations that dates back millions of years. These ancient lineages or "tribes" manifest themselves in the structure of the whale's mitochondrial DNA in the form of different haplotypes. Southern right whales show a remarkable diversity of mitochondrial haplotypes, at least several dozen different groups among females. In contrast, there are just five unique haplotypes in today's North Atlantic population. One of them is only present in just four males and will die out with them.

When McLeod analyzed the sixteenth-century bone discovered beneath the *San Juan,* she found a sixth haplotype not found in the modern-day population, evidence that there might have been a group of right whales that once visited the Strait of Belle Isle to feed and were killed off by the Basques. This possibility is intriguing because the loss of this other mitochondrial haplotype might represent a loss of knowledge of a historic feeding ground, helping explain why right whales are not seen in the strait today. The Bay of Fundy tribe, on the other hand, may have survived the whaling era simply because the waters there were considered by whalers to be too dangerous to hunt in.

The ability to analyze the mitochondrial DNA of right whales has shown scientists that whalers weren't just killing individual animals: the exploitation of North Atlantic right whales in their historical range—from the west Saharan coast, Azores, Bay of Biscay, western British Isles, and the Norwegian Sea—likely resulted in the extinction of a right whale culture passed through females. This realization makes the outlier whales like 1334 who are seen every now and then in strange places that much more fascinating.

"One of the old names for the North Atlantic right whale is *Noordkaper,* referring to the North Cape of Norway," said Bob Kenney. "One of our males, Porter, showed up there. We still get the oddball sighting off Greenland and Iceland and some in the Gulf of St. Lawrence." Could these outlier whales be retracing ancient migrations in search of food, revisiting places that are the right whale's historical habitat, faint memories of which are inscribed in their DNA? When the whales from the Bay of Fundy disappeared in 2013, is this how they found another place to feed, and is it vulnerable to climate change too?

In late 2014, Brad White showed me his laboratory's analysis of 1334's DNA. As expected, 1334's haplotype placed her in the "B" group of females that is composed of non-Fundy females. Her nuclear DNA revealed the paternity for two of her calves. One of the calves was fathered by 1055, a whale first seen in 1979 and since then most often sighted in New England's Great South Channel in late May. The second calf was the one Katie Jackson biopsied in 2013; its father was 1513, a whale called Crest for the stripe of white scarring on his tail that looked like a strip of toothpaste. Crest was first seen in 1985 in Cape Cod Bay and the next year had been spotted in Roseway Basin, an area south of Nova Scotia, during a record week when New England Aquarium researchers documented some seventy whales in the area. Mothers and calves are rarely spotted in Roseway; it's almost all males (though during the 1990s even they appeared to abandon the area because of a lack of food abundance). In recent years, researchers have found Crest most often migrating through the Great South Channel in the late spring on his way north. But where did he cross paths with 1334 to mate?

The greatest clue may be where Crest has appeared in the winter months, the estimated conception period for females, and from 2008 to 2010, Crest showed up in the Gulf of Maine. Right whale researchers were in agreement that this is most likely the primary mating ground for the species, and in 2013 seven scientists including Brad White published data from six years of aerial surveys in the journal *Endangered Species Research* that showed around half of the entire population was appearing in the Gulf of Maine between November and January. But as Bob Kenney described it to me, like

everything else with this perverse species, once researchers believe something about the whales, they go and do something different; since then it's
been hard to find any whales in the Gulf of Maine during the winter.

Where 1334 and her non-Fundy brethren feed during the summer
months is another ongoing mystery. Some researchers think the other feeding ground might be in the Gulf of St. Lawrence. For the last few years,
the Right Whale Consortium has launched an outreach campaign in Prince
Edward Island, Nova Scotia, New Brunswick, Quebec, and Newfoundland,
handing out pamphlets to over 300 wharves, Canadian Coast Guard ves-
sels, ferries, and whale watching companies, asking for information about
right whale sightings. Other researchers are analyzing data from acoustic
monitoring devices in the Scotian Shelf as well as the Cape Farewell grounds
south of Iceland and east of Greenland to see if they record right whale vocalizations. If they get a clue of where to look, they might start following in
the steps of scientists in Argentina who have begun using satellites to count
the southern population. I found it incredible that animals so large could be
so difficult to find that sometimes it takes a telescope the size of a satellite up
in space to see them. As much as I want to know where 1334 goes, I cheered
her elusiveness and hoped that the ocean is still big enough for her to escape
the forces threatening her kind.

5

FREEZING CROWS

Corvus hawaiiensis

oah built a boat to save the world's biodiversity; today, scientists build freezers. In the underbelly of the American Museum of Natural History, I met Julie Feinstein outside her laboratory, a place called the Ambrose Monell Cryo Collection. By my estimate we were somewhere beneath the Hall of Minerals or Meteorites, but it was hard to tell for sure; we had taken a bewildering number of twists and turns through the museum's exhibits and down into storage facilities and through shipping rooms to get there.

The American Museum of Natural History, or AMNH, has been a treasured New York City institution since 1869. School children and tourists flock to the famous brontosaurus and blue whale exhibits, and it's easy to forget that the museum itself is one of the least impressive aspects of the museum's operations; it is also a prolific research institution with over 200 scientists. This aspect of the AMNH is carefully hidden from the public view behind camouflaged doors that lead to labyrinthine hallways and secret staff-only elevators. Indeed, once you start learning what really goes on inside, the AMNH starts to seem like a magician's top hat from which end-less rabbits are miraculously pulled. The ten-story Childs Frick Building in the courtyard of the museum, for example, has 40,000 square feet of storage space and is the largest collection of mammal and dinosaur fossils in the world, but you can't even see it from the street. The whole collection of the museum recently surpassed 33 million specimens, but 99 percent of these are stashed away behind the public exhibits at the museum.

Feinstein's laboratory is one of the hidden marvels of the AMNH. One of the world's largest cryogenically frozen tissue sample collections, it lives in a tiny basement in a western edge of the building, near the corner of Seventy-seventh Street and Columbus Avenue. On the way to it, you pass a 3,000-gallon liquid nitrogen tank, surrounded by eight-foot fences with six-inch metal spikes on top, which feeds into stainless steel vats inside the laboratory. In these vats are whales and birds and 87,000 other samples taken from around the planet and now kept frozen at minus 160 degrees Celsius.

Feinstein, the curator of this collection, has a master's degree in botany and a hobby writing books on urban wildlife. Every morning she takes the subway from Brooklyn and gets off at Fifty-ninth Street, walking the last mile to work so she can observe the birds, insects, and animals that inhabit midtown Manhattan and Central Park. Dressed casually in blue jeans and a pink button-up shirt, Feinstein sat with me in her orderly office to talk about her work and the value of the AMNH's frozen collection. "Many are priceless, irreplaceable specimens. They might come from places where it is politically difficult to collect. It's not free to travel around the world to get them," she said. "All of them are animals that died for science, so in a way, they are all really priceless." Freezing tissue samples is not an easy task, or at least doing so in a way that preserves the integrity of their DNA for posterity. "It's hard to store tissues because they are filled with entropy and disorder. They are cold and hard to handle. People store them in inappropriate ways that are undependable," she said.

It was difficult to imagine the treasures the steel canisters, sweat-ing with condensation on that hot summer day, held within them, and I asked her whether I could look inside one. Feinstein put on plastic goggles and thick rubber gloves to protect against contact with the vat. It was large enough in circumference to require two or three adults to hug, and tall enough that Feinstein had to step onto a small platform in order to reach the top and open its lid. When she did, a thick white fog spilled over the sides. Inside, the temperature was cold enough to freeze a piece of fruit and smash it like glass. Feinstein invited me to step onto the platform, warning me not to inhale the vapor too deeply. I saw what looked like a giant *Trivial Pursuit* pie with six sections, each of them holding nine metal racks. With a gloved hand, Feinstein turned the pie and pulled a rack out. It had thirteen white boxes stacked on top of each other and inside each box were 100 two-inch vials, all labeled with a barcode and serial number. Using forceps she picked a vial from a box at random and rattled it to show me a specimen that looked like a black-eyed pea. "This is number 110029," said Feinstein, reading the barcode. We walked to her office in the next room and she opened up the collection's database on a computer. "Here it is," she said. "110029 is a mosquito from the New York City Department of Health." She paused for a moment, searching her own memory. "I remember this, it's someone's PhD work."

Since it was built in 2001, around 10,000 samples have been added to the Ambrose Monell Cryo Collection each year and it has the capacity to hold millions. Every organism except for humans is part of the collection's purview. The samples are the foundation of the museum's efforts to map the evolutionary relationships among organisms through their genetic makeup, and each are cataloged in a virtual database by phylum; class; order; family; subfamily; genus; common name; and continent, country, or body of water. Many of the specimens are exceedingly rare. They were collected from far-flung locales around the world: highly endangered Channel Island foxes; sea nautilus from the South Pacific island of Vanuatu; leopard frogs from the Huachuca Mountains. Among the most strange and obscure samples are an extensive collection of nonhuman papillomavirus samples taken from camels, donkeys, and manatees. The life's work of lepidopterologist Dan Janzen—samples from forty years of butterfly collecting in Costa Rica—is housed in the collection. So is the entire stock of tissue samples collected by the US National Park Service, including California condors, spotted owls, and Karner blue butterflies. It seemed comical that in a collection whose mission is to be the largest and most comprehensive collection of genetic diversity of life on the planet, Feinstein had randomly picked a sample that might have been plucked from a puddle not far from where we were sitting. Of course, it hardly mattered in a visual sense—the vials look identical except for what's inside.

Feinstein's job at the Cryo Collection is to impose order on a vast amount of data associated with the samples, making sure that each one is properly cataloged and available for any scientist around the world who might request them for research. She's like the librarian of a lending library, if librarians were expert molecular biologists whose collections are filled with irreplaceable books that would be unintelligible and meaningless if cataloged or preserved incorrectly. This is a job that requires tremendous capacity and patience for detail. She takes raw samples that are often collected under difficult conditions in the field and makes them conform to laboratory standards designed to last hundreds of years into the future. It's a process that she describes as rife with challenges, the first being that idiosyncratic scientists gather the samples. She grimaced when she related the time a biologist dropped off a black trash bag filled with irreplaceable herbs from the mountains of Mexico and a handful of Xeroxed field notes; it took a year and a half to catalog the 850 samples inside the bag.

Under Feinstein's stewardship, the museum's Cryo Collection has become a highly acclaimed model of genetic resource banking. When I met her, she had just returned from the first meeting of the Global Genome Biodiversity Network, an international consortium of frozen biodiversity repositories. During the conference some German documentary filmmakers had requested an interview. "You're a hero," they told her. I heard others praise Feinstein's leadership in the field of genetic banking. She is unique in that she has managed to master two very different disciplines—data management and molecular systematics. In London, the participants in the meetings had focused on advancing the idea of creating a single, globally accessible database of tissue samples. "It's the most boring subject in the world, listening to the database talk," said Feinstein. But there is an unmistakable strain of missionary urgency in the act of genetic banking. In an era of shrinking species populations and extinction threats, frozen tissue collections are increasingly seen as the best way to stem the tide of another kind of threat, what is sometimes called the secret extinction, the largely unseen and unrecorded loss of genetic variation within species. "It's a job with a genuinely important responsibility on a sort of global scale," said Feinstein. "You know, saving the planet every day."

*   *   *

For centuries, humans have collected the living things around us, stuffing them, sticking them with pins, and squirreling them away to feed the curiosity of scientists and the public in an attempt to try to understand the diversity of life on the planet. By 2050, the number of specimens preserved in the 6,500 or so natural history collections around the world are projected to increase by 500 percent from 2.4 billion. Around three decades ago, scientists began hoarding on a different scale, not specimens but genes. As the field of genomics—the study of DNA for understanding evolutionary biology, taxonomy, biochemistry, population genetics, and management of captive populations of animals—grew, scientists found themselves needing to maintain frozen tissues for research. But they started off doing so in very haphazard ways. "Lots of individual scientists at individual academic institutions would do research and their samples would end up who knows where," said George Amato, the director of the AMNH's Sackler Institute for Comparative Genomics, whose auspices the Cryo Collection falls under. "In a freezer in the back of their lab and then the freezer would break down, or they would retire. The countless amounts of materials that have disappeared . . . God." The Cryo Collection was designed to centralize and improve the potential scientific value of these types of samples—particularly from curators within the museum but from many others as well—and since then the number of repositories like it has ballooned. In 2011, the

Smithsonian Institution began building a new facility with the capacity to preserve 4.2 million specimens. The International Barcode of Life project is a consortium of genetic repositories whose goal is to create 5 million barcode records from the DNA of 500,000 species. The Genome 10K Project is collecting tissue and DNA samples from 17,000 species in order to sequence 10,000 genomes for analysis; the announcement of the initiative called it the biggest scientific study of molecular evolution ever proposed. "The ability to examine thousands of genetic markers with relative ease will make it possible to answer many important questions in conservation that have been intractable until now," said the founders. In Australia, the Western Plains Zoo in Dubbo has frozen 70 billion coral sperm and 22 billion embryonic cells in liquid nitrogen, which might be used to raise species of coral larvae under threat in the Great Barrier Reef.

Science funding and research undergo a constant shift of emphasis, a kind of trending that mimics fashion in its finicky and cyclical nature. Today, the cryopreservation of the world's biodiversity is in considerable fashion, and the enthusiasm for preserving genetic materials is akin to the nineteenth-century zeal for herbariums, zoos, and natural history museums themselves.

One of the first individuals to recognize that the world's genetic diversity is in need of saving was an Austrian-born geneticist and plant breeder, Otto Frankel. Born in 1900, Frankel was a young communist who wanted to dedicate himself to solving problems of hunger and chose to study agriculture at university. In the 1960s, he became the so-called prophet of genetic resources conservation, starting with his involvement in the International Biological Program, an effort to coordinate large-scale ecological studies and priorities, which held its first general assembly in Paris in 1964. A few years later, he helped to organize a conference on "The Exploration, Utilization and Conservation of Plant Genetic Resources," considered a milestone for the genetic resources movement. Otto recognized there was a growing loss of genetic diversity among plant species and believed institutions needed to respond with long-term seed storage, computerized data cataloging, and the creation of a global network of genetic resource centers. Most significantly, he argued that humankind's impact on genetic diversity was on so great a scale that we had "acquired evolutionary responsibility and must develop an 'evolutionary ethic.'" In 1974 Frankel went to Berkeley, California, for the International Congress of Genetics and presented his paper "Genetic Conservation: Our Evolutionary Responsibility," which according to subsequent leaders in the field of conservation biology was groundbreaking in its presentation of a conceptual and moral agenda for conservation. An evolutionary ethic, said Frankel, is one in which civilized man recognizes the continued existence and evolution of other species as integral to his own existence.

> Neither our pre-agricultural ancestor, nor the peasant farmer who succeeded him had cause for concern beyond the next meal or the next crop, the former because he used a pool of great species diversity, the latter a pool of self-renewing intraspecific diversity. This came to an end with the advent of scientific selection. Today's concern is with preserving and broadening the genetic base. The time perspective for gene pool conservation might be the next 50 or 100 years—which is merely an acknowledgement of the unparalleled technological transience of our age; we cannot foresee even what kinds of crops will be used at that time. For wildlife conservation the position is altogether different. Concern for its preservation is new, a consequence of our destructive age. Nature conservation is fighting for reserves and for legal recognition. The sights often are set for the short term, although perpetuity is its ultimate objective. Genetic wildlife conservation makes sense only in terms of an evolutionary scale. Its sights must reach into the distant future.

A year after Frankel delivered this message in Berkeley, a molecular biologist by the name of Oliver Ryder at the San Diego Zoo began collecting and freezing tissue samples of wildlife for the institution's "Frozen Zoo." The primary purpose of the collection was conservation of rare and endangered animals, and like Frankel described, it sets its sights into the distant future: it was an ark built for genes. The need for these collections was growing in the minds of biologists in the 1970s. "There is a benefit in maintaining genetic diversity not only among species but within species," said Norman Myers in 1976. "I believe that we should keep as many options open as possible until, through research, we can reduce the areas of uncertainty. A principal conservation need is to set aside sufficient representative examples of biotic provinces to extend protection to entire communities of species."

At the Frozen Zoo, the tissue samples from genetically valuable individuals were not the only thing put on ice. The zoo cryopreserved sperm, eggs, and embryos from rare individuals to be used in an array of assisted reproductive techniques, a process that conservation biologist Robert Lacy has described as stopping or freezing evolution: "Cryopreservation of gametes and embryos can provide a powerful tool for slowing evolution by allowing use of long-dead donors as the genetic parents of future generations." In San Diego, biologists used cryopreserved sperm to artificially inseminate Chinese pheasants that then successfully hatched new chicks, and harvested oocytes and sperm from dead southern white rhinos to use for in vitro fertilizations. Since its inception, the Frozen Zoo has gathered tissue samples from more than 10,000 individual animals representing over 1,000 species, and as scientists create ever more sophisticated technologies to use in assisted reproduction and cloning, the collection has become both metaphorically and literally a freezer from which scientists hope entire animals can be preserved and resurrected.

The Frozen Zoo, the Cryo Collection, and other gene repository initiatives around the globe are designed to last far into the future, a time when we can hardly predict what seemingly miraculous technology scientists will have for their holdings. In some cases, the samples are like snapshots of moments in history that might otherwise be lost: frozen water samples of a wetland might reveal microbes that are critical for restoring the ecosystem a hundred years from now. This unknown potential is part of what makes the samples priceless and ignites the imagination, giving scientists and the public a sense of "what if" exhilaration and hope. "There is a latent potential in these collections," said Joanna Radin, a scholar of the history of medicine and science at Yale University. "Maybe what's actually going to be useful in the future is the microbial diversity on the samples, maybe that's going to be the true contribution. The potential value of the collection isn't known, and when they talk about these collections, people will say, 'We don't know exactly what they are good for, but they will be good for something.'" The tissues within the Cryo Collection are maintained at a much colder temperature than is needed just to maintain DNA for research, thereby keeping all the options for the samples open, said Radin. Her research focuses on the history of the idea of freezing life, starting with the invention of cold storage in the twentieth century that allowed anthropologists and geneticists to transport human blood samples from the field to laboratories. Cryopreservation helped make possible the notion of genetic salvage, not just of wildlife and plants, as Otto Frankel emphasized, but also races and ethnicities that were seen to be in danger of extinction. Indeed, one of the aims of the International Biological Program was saving samples of "unpolluted" nature, including blood and tissue, from so-called primitive peoples around the world. These scientists collected in a state of anticipation. They froze blood in order to save the genetic secrets within it for a time in the future when science could reveal them but the people themselves might have disappeared. According to Radin, these samples, like those in frozen biodiversity collections and zoos, are examples of "latent life," a state in which a biological substance is neither fully dead nor alive.

Over a few months, I made several visits to the Cryo, and after I left the laboratory I'd visit the upper floors of the museum to look at the exhibits. Standing in front of the fossils of extinct species like ambiguous dogs and ruminant horses, I would try to focus my thoughts on the transience of so many life-forms and the significance of the phenomenon of extinction. In the Wallace Wing of Mammals and Their Extinct Relatives, I looked at dimly lit dioramas of African mammals and it struck me that the instinct to freeze these scenes of wildlife in time wasn't so different from the collection in the basement beneath me, where DNA sat frozen in vials. The visionary behind the Hall of African Mammals was Carl Akeley, a biologist, big-game hunter, taxidermist, and photographer whose obsession was Africa. Akeley believed that he had to preserve the continent's charismatic wildlife, the giraffes, lions, and rhinos, in order to draw attention to the disappearance of the African landscape. "The old conditions, the story of which we want to tell, are now gone, and in another decade the men who knew them will all be gone," he wrote in 1926. Akeley was particularly taken with the gorillas of the Virunga Mountains in then Belgian Congo, a species scientists had just discovered in 1902. Akeley felt a strong kinship toward the gorillas, so much so that when he shot some of them to bring back to New York City, his conscience was troubled by the experience. "It took all one's scientific ardor to keep from feeling like a murderer," wrote Akeley afterward. "Of the two, I was the savager and aggressor."

Akeley considered himself an artist when it came to preserving his specimens. He didn't stuff the animal with straw as was the standard practice of the day; he mounted the skeleton of the gorillas in a desired pose and built each muscle and tendon in clay, so that when he finally placed the skin over it, the animal looked as natural as it would in a photograph taken in the wild. To the modern eye, gorged on IMAX, Akeley's dioramas seem a little dispirited and artificial. But at the time they were revealed to the public, they were considered the height of realism. In the fifty years following his death, Akeley's fears about Africa's wildlife were realized. The mountain gorilla population declined from thousands to between 250 and 280 individuals. Nearly 45 percent of land in sub-Saharan Africa is agricultural. But Akeley's animals are still frozen in time, preserved for future generations to look at. The Cryo Collection collapses space and time into two-inch vials—I don't know if what I'm looking at is a whale or a mosquito—but it isn't so different: both are an acknowledgement that time is slipping by, sometimes so fast we have to hit the pause button before things disappear. In a way, freezing animals is a concession that we're not sure how else to save them. "I think fundamentally freezing is a technology of deferral," said Radin. "It's for deferring action to the future. And the future may be a future that never actually arrives. The value of these collections is that they allow us to say look, we know there is a problem and science has been a huge part of creating that problem in terms of the byproducts of industrialization. We don't know what we are going to do, but in the future, other people will have

better answers and so we owe it to future generations to preserve some sub-
strate of the world we are destroying."

I asked Radin if she thought it was possible that frozen collections are
a sign that the conservation movement is literally conceding ground in
the struggle to preserve nature. Have we gone from trying to save landscapes to trying to save DNA? "I would be loath to say that one replaces
the other," said Radin. "Inevitably, there are certain choices that get made
in any scientific project. There's an argument that can be made about the
increasing reductive nature of this process. First we tried to preserve entire
habitats, then the animals and their behaviors—even if not in free-rang-
ing habitats. Now there is a sense that if we preserve the DNA then we have
everything we need to know about them." But, Radin said, after decades
of enthusiasm for the idea that preserving DNA saves essential data about
species, today this notion is coming under scrutiny even in medical and
scientific circles.

Some environmental ethicists have been ringing the alarm bell for a
long time. One of the most compelling and provocative antidotes to the
tendency toward genetic reductionism comes from Holmes Rolston III.
Rolston's theory of *telos* in species, that they have an ultimate object or aim
that is only fulfilled in kinship with other organisms, is unique but potent
to consider in the context of genetic banking. He describes these ideas in his
book *Genes, Genesis, and God,* which he was inspired to write after reading
Richard Dawkins's *The Selfish Gene,* a book that Rolston believed warped
and exaggerated the role of genes in biology. For Rolston, ecosystems—
whether it is a mammalian uterus or a forest—is as ultimate a truth as a
gene. An organism uses its genes to act for its own sake or to protect its intrinsic value. But an organism does not *just* act for itself; it acts to maintain
a historical trajectory between itself and its species as a whole—a telos that
can only be expressed in relationship to others.

The life that the organismic individual has is something passing
through the individual as much as something it intrinsically possesses.
All such selves have their identity in kinship with others, not on their

own. This individual and familial identity is placed in a species line that must be historically maintained in the death and regeneration process, with both information stored at the genotypic level. A species is another level of biological identity reasserted genetically over time: sequoia-sequoia-sequoia, bee-bee-bee. Identity need not attach solely to the centered or modular organism; it can persist as a discrete pattern over time. The individual is subordinate to the species, not the other way around. The genetic set, in which is coded the telos, is as evidently the property of the species as of the individual through which it passes.

When I talked to Rolston from his home in Colorado, he put it a little more simply: if you plop a gene down on the moon, absolutely nothing happens. A gene without a world to function in is unintelligible, it is what it is only in interaction with its environment. There is no telos without relationship. "People said that genes called the shots, that genes are ultimately what's in control of how an organism is built and what it does and so forth," he told me. "Since that time, a great deal of research and thinking has been done in epigenetics to the effect that yes, the genes are there and you can't do without them, but they interact *with* the environment. It's a two-way deal. It's not just genes producing an organism that succeeds or fails. The genes are in control of the organism, but immediately that organism is going to interact with the environment. It finds itself in hot or cold or wet or dry or something is coming around that might eat it. What the organism does is it uses the genes as a resource tip, 'Hey, I need some more of this enzyme because it's getting drier.' And it cuts on and off the genes that produce it. There is a two-way interaction between environment and genes. Every biologist in the world will say it's more complicated than we thought."

*   *   *

Within the freezers of the Frozen Zoo are the cells of the 'alalā, or Hawaiian crow, that once inhabited the forests of the Hawaiian Islands. The tissues of *Corvus hawaiiensis* are generally taken from the eyes or trachea of birds that have died. These are cut into small pieces and mixed with enzymes and cultures that differentiate and then feed the cells. Over the course of a month the cells grow and multiply many times over, and eventually a cryoprotectant is added that prevents their membranes from breaking when they freeze. The concoction is split into one-milliliter vials and placed into a freezer that gradually drops to eighty degrees below zero, at which point the cell lines are ready to be frozen in liquid nitrogen at temperatures of minus 196 degrees Celsius. "Cells can be kept frozen for many years in this suspended but living state; nobody knows exactly how long because the technology has only been around for a few decades," wrote Andrea Johnson, a research technician at the San Diego Zoo, in a description of the freezing process. In their frozen state, the cells of the 'alalā are full of possible applications for conservation. A few years ago, the Genetics Lab at the San Diego Zoo Institute for Conservation Research examined the mitochondrial DNA from the ten founders of the only population of 'alalās alive today. Just two of the genomes were distinct from each other. These sorts of findings give biologists a glimpse into the birds' evolutionary potential and ability to respond to environmental changes and diseases. It can help them understand and optimize the breeding potential of the remaining birds, and it might one day be possible to use the frozen cells to create sperm and embryos to make new crows and add them to the population. "One of the rewarding things about working with the Frozen Zoo is knowing that it will be a resource for future conservationists whose technology and goals we can only imagine," wrote Johnson.

The word 'alalā has a multitude of meanings and connotations within Hawaiian culture, where the bird has a spiritual and symbolic place. Some say the word comes from *ala* (to rise up) and *lā* (the sun), a reference to the bird's habit of filling the forest with its voice in the morning, while others say the word refers to the sound a child makes. At the court of Kamehameha in the eighteenth century, the 'alalā were a group of orators used by the king to deliver news in poetic form or songs, or during wartime to communicate commands to warriors.

'Alalās are large and have midnight black feathers and blue eyes that turn to brown as they mature. They were abundant in Hawaii's montane forests and mountain slopes, places where 'ohi'a and koa trees grew, and fruits and seeds as well as insects and small vermin were plentiful. The forest canopy helped the crows protect themselves and their young from the 'io, a Hawaiian hawk that is the only forest bird larger than them. Like other corvids, Hawaiian crows are intelligent and emotional. They have been observed using twigs as tools to get food and are known to have lived as long as eighteen years in the wild. They are monogamous, forming long-standing bonds with another bird and building nests together in early spring before raising a brood of one or two chicks each year. Their vocalizations are raucous, ranging from howling to growling to muttering. In the Hawaiian language, one of the meanings of 'alalā is "to bawl, bleat, squeal, cry." In the late 1800s, the archeologist Henry W. Henshaw wrote, "It would be difficult to imagine a bird differing more in disposition from the common American crow than the Hawaiian 'alalā. The bird, instead of being wary and shy, seems to have not the slightest fear of man, and when it espies an intruder in the woods is more likely than not to fly to meet him and greet his presence with a few loud caws. He will even follow the stranger's steps through the woods, taking short flights from tree to tree, the better to observe him and gain an idea of his character and purpose."

The 'alalā's struggle to survive most likely began when the Polynesians arrived in Hawaii some 1,500 years ago, bringing "canoe species" such as pigs and deer that competed with the birds for food. In the 1800s this pressure was compounded by the arrival of colonists who brought rats, mongoose, and cats to the islands. Hawaiian forests became fragmented from ranching and logging, and diseases such as avian pox and malaria increased. By 1950 'alalās were living in small pockets of what undisturbed forest was left. By 1985 their survival had become dire: biologists believed just five to fifteen birds still lived in the wild, most of them on the McCandless Ranch, a privately owned tract of land in the Kona District. Under this threat of extinction, the US Fish and Wildlife Service took a handful of wild birds into captivity, eventually moving them into an endangered-breeding facility at

Pōhakuloa on Hawaii, where efforts to save the Hawaiian goose, Hawaiian duck, and Laysan duck were already under way. Of this founding population, only four pairs of crows produced offspring.

During the late 1980s and early 1990s, when the wild population of crows was ailing so badly and the captive breeding program was struggling, the process of capturing more wild birds for breeding was contentious, pitting private landowners against the US Department of Interior, Sierra Club Legal Defense Fund, National Audubon Society, and eminent conservation biologists. Cynnie Salley, whose family began amassing the 60,000-acre McCandless Ranch in 1915, was intensely protective of the flock of crows living on her land. The biologists coming to study the birds, she believed, were another disruptive presence in the crows' environment and, just like the government's captive breeding program, acted without the species' best interests—survival in the wild—at heart. This story is described by journalist Mark Jerome Walters in his captivating 2006 book on the conservation history of the crows, *Seeking the Sacred Raven: Politics and Extinction on a Hawaiian Island*. Salley once responded to the state's governor, who in 1991 begged her to open her property to experts to study the 'alalā, by writing, "The arrogance implied, through you, by the wildlife biologists, that their approach is the only salvation of the 'alalā, is repugnant and in error."

The fight over the 'alalā was a symbolic struggle in a place where sovereignty, colonialism, and federal control had long been potent forces in Hawaiian history. It took a lawsuit and a scientific committee formed at the behest of the National Research Council to solve the impasse between the government and conservationists on one side and Salley, symbolically representing native Hawaiians and land owners, on the other. In 1993, the parties signed an agreement that allowed biologists to take eggs hatched by the wild flock instead of adult 'alalā. These eggs would be hatched in captivity, and some of the chicks would be released while others would be added to the captive population to add genetic variety. But between 1993 and 1998, twenty-one of the twenty-seven birds raised in captivity and then released in the wild died.

In 1996 the last fertilized egg was taken from a wild nest by biologists. One of them, Alan Lieberman, a program director at the San Diego Zoo's Hawaii Endangered Bird Conservation Program, has written that in hindsight, this single egg was of monumental significance for the captive breeding program. It could have been unfertilized. It could not have hatched. Luckily, it was and it did. When a male chick was born on June 9, biologists called it Oli, meaning "ritual chant." Oli wasn't a particularly prolific breeder; he went on to produce just six offspring. But his genes proved to more than make up for his lack of enthusiasm. Oli's chicks were fertile and they begat other fertile chicks, and his line extended through the generations. In 2002 the last two 'alalās in the wild disappeared, and no other 'alalās have been seen in the forest since. The breeding program now contained the species' entire genetic pool; 47 of the approximately 114 'alalās in the captive population are descended from Oli's genetic line. These 'alalās have become prolific enough breeders that biologists believe they might be able to release several crows into a managed forest reserve, where they could be closely monitored and semidependent on humans for food and veterinary care. The species, Lieberman says, is a key, perhaps *the* key, to healthy Hawaiian forests, needed to disperse the seeds from a variety of native plants and trees including the loulu palm and halapepe. The ho'awa, a shrub with walnut-like fruit, is dependent on 'alalās for its dispersal and can only germinate after passing through the bird's digestive system. Already, the absence of 'alalās has changed the composition of plants and trees in the forests they once lived in. Without Oli, the tantalizing possibility that these ecosystems and codependent relationships between species could be restored one day might have been impossible.

Much like the single fertilized egg found in 1996, the cell lines of 'alalās preserved within the Frozen Zoo might prove essential to the 'alalā's future survival—no one knows for sure. But preserving the bird's DNA cannot prevent another kind of 'alalā extinction that is undoubtedly already taking place, that of the bird's culture. Hawaiian crows are complex creatures that demonstrate a capacity for learning and social interaction similar to primates and dolphins. I learned about crow culture in the work of Thom van

Dooren, a senior lecturer in environmental humanities at the University of New South Wales in Sydney, Australia, and a leading academic in the emerging field of extinction studies with a particular focus on birds. Van Dooren's research, as he has said, brings biology and ecology into conversation with philosophy. His book *Flight Ways: Life and Loss at the Edge of Extinction* examines the fates of albatrosses, Indian vultures, and whooping cranes in the twenty-first century, and he writes about these species as full-fleshed characters in the midst of an unfolding tragedy, as millions of years of evolution come to an end in our lifetimes.

From Germany, where he is a visiting scholar at the Rachel Carson Center for Environment and Society, van Dooren told me he has always been fascinated by crows and ravens. He spent weeks at the ʻalalā captive breeding center in Hawaii, observing the birds and talking to caretakers, biologists, and local Hawaiians. Seeing the last ʻalalās in the confinement of an aviary is bittersweet, said van Dooren. His first impression was that the birds moved through their long cages much like the nineteenth-century ornithologists described them in the forest; half jumping and half floating through the branches, wanting to get a closer look at who had entered their home. "I definitely felt a mixture of happiness and sorrow when I visited the crows," he told me. "Crows in captivity are incredibly sad because they are such adaptable, clever social animals."

The question of whether ʻalalās have maintained the behaviors of forest-dwelling birds is central to the conservation effort, explained van Dooren. Unlike some bird species, ʻalalās are not hardwired from birth for particular behaviors. Instead, they learn how to be a crow, so to speak, from their parents in the year after hatching and when the juvenile birds join the larger flock, which can contain multiple generations of the young bird's family. For over two decades, the eggs of captive ʻalalās have been pulled from the parents' nest and hatched in incubators to ensure their survival. Until 2013, when the first female bird was allowed to hatch and raise her own eggs, all ʻalalās in existence were artificially incubated, then hatched and raised by humans. There is evidence that the birds' culture has already changed in substantial ways as a result; behaviors particular to the ʻalalā, once passed from generation to generation, have disappeared. The birds' repertoire of vocalizations has diminished. When a release of captive-bred 'alalā was attempted in the 1990s, the crows appeared to no longer know how to avoid the 'io, the Hawaiian hawks they once banded together against. The crows are habituated to humans and no longer forage for food. These lost behaviors could prove detrimental to their ability to survive in the wild. Van Dooren spoke with local Hawaiians about the 'alalā, including Cynnie Salley, who fought so bitterly to keep the last wild 'alalā on her land. Salley told van Dooren she believes the captive breeding program has changed the crows profoundly, so much so that they represent a different species.

> They were kind of like the kings and queens of the forest. They chased the hawks and the hawks had a healthy respect for them. As a matter of fact, it took four or five years of releasing young birds before the hawks realized that these were different than the ones that used to chase them around and that they had fair game. . . . All of those birds that were originally wild are now gone. All of the birds there [at Keauhou Bird Conservation Center] have been raised by puppets. So I truly feel that whatever happens in the forest now with these birds, it's a different species. . . . Whatever they release now is really starting at evolutionary ground zero. They're going to have to relearn everything—including calls. . . . So, from their language on up they're going to have a huge learning curve.

For van Dooren the question of species authenticity and identity are particularly befuddling and potentially unhelpful when it comes to corvids, whose intelligence have allowed them to adapt to human civilization over thousands of years in uncanny ways. Today in Japan, for example, jungle crows use moving cars to open nuts and then retrieve their food when red lights stop traffic. Crows in North America have become successful scavengers of human garbage. Maybe those behaviors are what it means to be an authentic crow in the twenty-first century. Van Dooren told me he thinks it's better to be flexible about what an 'alalā should and shouldn't do. Even if they don't behave exactly like their ancestors—feeding from Dumpsters rather than trees, for instance—they might still be considered 'alalā. But there are reasons, he believes, to value cultural continuity from wild crows to captive crows to released crows. "Conserving species is, at least in part, about holding on to evolved (even if still evolving) 'ways of life,'" he has written. "Ways of being in the world are at stake here, complex more-than-human cultures." From van Dooren's perspective, rooted in both philosophy and anthropology, the notion that extinction is the death of a species is a far from adequate understanding of what is happening to animals like the 'alalā today. Extinction is the slow unraveling of "delicately interwoven ways of life," generations of learning interrupted and lost from the world.

For these reasons, van Dooren believes it is deeply problematic to call genetic banking of tissues conservation at all. "The idea that having isolated the genome we have somehow captured the essence of an organism and species is incredibly reductive," he told me. "The notion that behavior is genetically determined is not how developmental processes work. Sadly, this comes up again and again. I did my PhD on intellectual property and plant resources. People are patenting gene sequences as if a genome is a blueprint of an organism." In an essay for a forthcoming book on cryogenics, van Dooren challenges the notion that banking, and embracing the implicitly hopeful future this practice represents, is unproblematically good: "All cryo-technologies of endangered species conservation have in common the fact that, to a greater or lesser extent, they fail to *conserve* what many people want to hold on to. There is a kind of reductivism underlying these *ex situ* approaches to conservation in which 'accessions'—be they living birds, seed or DNA samples—must be held outside of the vast web of relationship that have given rise to and sustained them." We can't put culture in liquid nitrogen, the same way we can't bank the forest the 'alalās come from. No one would say that freezing the DNA of humans preserves what makes us human.

But van Dooren doesn't go so far as to say that genetic banking should be abandoned. The fact is that much like captive breeding facilities, they are approaches of last resort, and without them 'alalā might not survive at all. No doubt, this is why the experience of seeing 'alalās in captivity is so bittersweet: there is hope in the fact that they are still living—there is still, as the Hawaiian phrase, goes, "*Ke nae iki nei no,*" some breath remaining, and sadness that it is in such a diminished form.

In *Flight Ways,* van Dooren writes that humans might be able to learn something about mourning from crows themselves. The notion that a species could teach humans a behavior that is considered fundamental to our humanness—the ability to grieve—struck me as radical. But crows have been observed mourning, calling to, and revisiting their dead, and even avoiding places where other birds have died. As van Dooren writes, "If the death of a single crow signals 'here lies danger'—a danger significant enough to avoid a place for years, to alter flight ways and daily foraging routes—then what must the death of a whole species of crow, alongside a host of others at this time, communicate to any sentient and attentive observer? How could these extinctions not announce *our* need to find new flight ways, new modes of living in a fragile and changing world?"

Later, I learned that in Hawaii, the wail of women that have lost a loved one is also called an 'alalā.

＊　＊　＊

It would be perverse to criticize a doctor for treating a patient with a bullet wound rather than spending his time lobbying for gun control legislation. For the same reason, it's difficult to criticize proponents of gene banking for not focusing on the preservation of species in their environment. I started to think of conservation biologists and geneticists like ER doctors; they operate in an emergency framework, triaging the loss of biodiversity in a time of calamity. Faced with the threat of extinction, they are forced to make quick decisions with limited funds to save species, often by preserving what remains of a limited genetic pool before it disappears. Surely these frozen snapshots of life stored within vials will be needed in the future, to remember, understand, and possibly even re-create what is inside. Who is to say that in an age of accelerating extinctions, genetic banking isn't some of the most important work taking place in conservation? We can't predict what disasters are imminent and whether gene banks will be all that withstands the coming floods of ecological crisis, giving us the ability to one day resurrect landscapes that might have been lost forever.

This apocalyptic logic permeates genetic banking initiatives around the world. In Norway, a vault has been carved into the ice shelf near the North Pole, a place for seeds representing the agricultural biodiversity and food supply of the world to be kept safe in the event of nuclear war or natural disaster. Called the Svalbard Global Seed Vault and located in the Svalbard archipelago, the site was chosen in part because it is high enough above sea level to remain dry even if the polar ice caps melt from the effects of global warming. The Frozen Ark, a consortium started by British scientists in 2006, rationalizes its efforts to cryo-bank samples from 10,000 species by 2015 by stating that habitats such as coral reefs, ice caps, and the majority of the rainforests are no longer likely to be adequately conserved. Harvard entomologist and conservation leader Edward O. Wilson has said that the preservation of every scrap of biodiversity is an ethical imperative until humanity understands its value, a quote the Frozen Ark cites in its mission statement.

Yet as the ʻalalā shows, gene banks preserve a miserly notion of what animates a species and its full potential. It deep-freezes evolution rather than allowing the wild, dynamic process of natural selection to unfold within a landscape. Nonetheless, gene banks are becoming a line of defense in the struggle to conserve species. The anthropologist Tracey Heatherington suggests genetic banking might seem so appealing today because it suspends political realities and moral conundrums for a future time and place. "With its biblical reference to imminent world disaster and its transcendent faith in technoscientific interventions, the Frozen Ark project reflects the moral discourses of global environmental movements that, after the multiple failures of the initiatives envisioned at Rio, are now permeated with urgency and irony," wrote Heatherington in an essay called "From Ecocide to Genocide: Can Technoscience Save the Wild?" Heatherington, a professor at the University of Wisconsin-Milwaukee, told me that the act of genetic banking reveals a paradox of modern science: as we seek to stem the loss of biodiversity through technological means, this very act compromises the authenticity of wild nature that consists of mutually embedded forms of both cultural and biological life. For all these reasons, Heatherington finds the biological essentialism and faith in technological solutions represented by genetic banking to be highly dubitable. "The idea that the behavioral profile of the animal is reproduced just on the basis of the genes is entirely based on this sense that animals have no learning process, no culture, that they don't learn and impart information to other species. That's just tremendously anthropocentric—that there are no species cultures or relationships beyond biology and that biology can be reduced to genes," she told me.

The environmental philosopher Bryan Norton has criticized the triage mentality among conservationists because it leads them to treat endangered species as merely a problem of protecting genetic diversity, and species to be regarded as a repository for a set of genes. In his book *Why Preserve Natural Variety?* Norton might have been describing the case of the 'alalā or Florida panthers or North Atlantic right whales when he wrote,

> Loss of genetic diversity is a manifestation of the deeper problem of decreasing biological diversity. As natural habitats are altered, converted, and simplified, many species suffer a decline in their number of independent populations. Attempting to protect genetic diversity through the protection of a few remnant populations ignores the most basic problem and will result only in a continual scramble to save individual species. A true solution would halt the tendency of more and more species to become so severely depleted that they require individual attention. If the deeper problems causing this tendency are not addressed, it can be expected that the effort to protect endangered species in remnant populations will become overwhelming.

Norton wrote these words in 1987, and since then the scramble has become more frantic and more urgent.

Even some conservation biologists doubt the possibility that genetic banking is truly relevant to the process of conserving species. Michael

Soulé, one of the founding fathers of conservation biology, challenged his colleagues at a symposium in San Diego in 2000 by saying that there is no example of a successful high-tech approach to conservation. Implicit in his challenge was a critique of a field that was enamored with genetic analysis and management of endangered populations of species as a conservation strategy. In the audience that day was George Amato, then director for conservation genetics for the Wildlife Conservation Society. Amato has helped define the field of conservation genetics through his prolific research and leadership at significant scientific and conservation institutions. Some fifteen years after Soulé put forth his challenge, Amato told me not much has changed; it's still difficult to point to a successful case of a high-tech approach to saving a species. Amato said this while sitting back in a chair at a small round table within the offices of the Cryo Collection, several floors beneath his office as director of the Sackler Institute for Comparative Genomics and just a few yards from the steel vats containing the collection's frozen tissue samples. For a grandfather, Amato is youthful and athletic looking, and he speaks with ease and candidness about the shortcomings of conservation genetics. "There are some really high-tech programs at a few zoos," said Amato upon reflection. "The criticism is they are not on the size and scale to do something significant. They tend to be these one-off exercises that get into the newspaper." In his writing, Amato has challenged conservation geneticists to be aware how their very efforts to mitigate extinction threats create a reductive paradigm, to "keep the endangered forest in sight as we genetically characterize every single tree." He believes the technological intervention into endangered animals can serve to further separate them from the environments that are intrinsic to their identity, and that scientists need to curb the press and public's fascination with genetic manipulation. But Amato told me the Cryo Collection and the implementation of an exemplary genetic banking system for future research is the biggest contribution he could possibly make to his field. "Human impact on the planet is happening at a pace far faster than our ability to internalize or comprehend it," said Amato. "We would be negligent not to, in a sensible way, get those specimens."

I had grown accustomed to conservationists' discomfort at framing their work in terms of ethical beliefs or personal feelings. If pushed to describe the moral roots of their convictions, they generally gave me utilitarian arguments for species preservation, along the lines that species are intrinsic to ecosystem health, which is in turn intrinsic to human survival. This may be true in the very long view, but as I found time and again the exact opposite is true in the short term: for people around the world, cutting a tree for fuel, damming a river for electricity, tilling the land for planting, or poaching an animal for money means survival. In order to succeed at their work, few conservationists navigate or even acknowledge this ethical relativity. To them the rightness of saving species is a kind of inviolable truth, even if others think it is a privileged opinion. Amato was one of the first and only conservationists who not only acknowledged this relativity in our conversation but was also comfortable with it. "I recently got into a fight with a group of people who live and breathe conservation biology as a religion," he said. "The reason *I* do this is I have to do something. I enjoy working in this realm. I feel like it's a worthwhile effort and all those things. But I'm not judging the value of my life based on whether or not there's going to be elephants twenty-five years from now. I'm not sure it's within my control." The possible extinction of elephants is not why Amato thinks you should care about extinction either. "I don't think it's *you* have to care about conservation because elephants will be extinct twenty-five years from now. It's about quality of life, it's what *kind* of planet do you want to live on, across the whole spectrum. In the suburbs, do you want to have salamanders and songbirds? Does it make a difference to know that someplace there are lions, or you can go to some place where you don't see all the impacts of humans?"

Amato's use of the word *religion* was revelatory for me. I was used to conservationists speaking about their work in scientific terms, but Amato was now describing conservation as something else entirely, not science but rather adherence to a set of preferences or beliefs. "[Conservation] is a complicated place," he continued. "It's a human ethical construct. If you wanted to be purely scientific about it, you have to ask: what is the scientific question? Is the world better with gorillas or without them? That's not science, it's values." The hard truth, said Amato, is that humans could survive in highly modified environments with less species for a very long time. Conservation biologists are scientists whose work, at its core, is an ongoing ethical debate about the future.

I have never seen an 'alalā in person. But when I thought about what Amato said, I found that I cared very much whether 'alalā are still living and breathing in the world, and that there is a glimmer of hope they could live in a forest again one day. Much less comforting was the knowledge that 'alalā tissue exists at the Frozen Zoo in San Diego. Is it conceivable that oblivion is somehow preferable to a freezer? These conflicted feelings might be rooted in romanticism, nostalgia, and my own particular aesthetic preferences. But the sadness is real nonetheless, and other people share those feelings. In *Seeking the Sacred Raven*, Mark Jerome Walters writes about Barbary Churchill Lee, a volunteer caretaker who worked with captive 'alalās from 1976 to 1981, and was fired for her controversial decision to bury the body of a female crow, 'Ele'u, that had died on her watch from bird malaria, rather than give the carcass to the government biologists. Walters describes how Lee brought the bird to state officials who culled some of its tissues; she then decided to take 'Ele'u back to the mountain where she had been captured, hiding the bird's remains, in a sense, from science. As Lee told Walters:

> I could not limit myself to thinking like a scientist. I had feelings! I will never forget driving up the mountain at Hualalai with 'Ele'u in a small box on my lap, tears streaming down my face. In my lap I held one of the last broken links between a species' existence and its extinction. There was this deep voice of Hawaiian history and belief in me, telling me that to bury 'Ele'u was the only thing to do. In the old days, warring groups hid the bones of their own for fear the enemies would dig them up and defile them by making fishhooks and other objects. To this day, no one knows where the bones of King Kamahameha are hidden—probably deep in a cave somewhere along the Kohala Coast. I didn't want the

enemies of 'Ele'u to have her bones, either, because they would defile them in their own way, leaving them in a freezer or dried on a shelf until they get thrown out. I know burying her wasn't technically the correct thing to do, but it was the right thing to do.

6

METAPHYSICAL RHINOS
Ceratotherium simum cottoni

n early 2008, stem cell researcher Jeanne Loring decided to take her staff on a field trip to the San Diego Zoo. Loring had recently been recruited by the Scripps Research Institute in La Jolla, California, and the visit was an opportunity to thank her staff for moving their laboratory. It was an unusual field trip for this group of scientists. For decades, Scripps has been on the leading edge of medical research, developing and testing treatments for leukemia, HIV/AIDS, and multiple sclerosis. Loring's research is in the field of regenerative medicine—how to engineer human cells to treat and cure neurological disease. Regarded as a pioneer, she was one of the first people to master the production of human embryonic stem cells in a laboratory. She is a passionate evangelist for her field, involved in landmark legislation and patents. Loring describes herself as a scientist who likes big questions, and stem cells, she says, let you ask them. The subtext of the trip to San Diego was one of these big questions: Could they employ stem cell technology for the purpose of wildlife conservation?

Loring knew about the Frozen Zoo, the collection of tissue samples from more than 1,000 species run by conservation geneticist Oliver Ryder. There had even been some talk between Loring and Ryder before about how embryonic stem cells might be used in conservation biology. "The trouble was, the technology didn't exist," said Loring. "It was just talk." Harvesting stem cells from the embryos of endangered animals was both logistically and ethically challenging.

In 2006, however, Japanese scientist Shinya Yamanaka published his method for reprogramming any mature living cell into a stem cell, creating what is called induced pluripotent stem cells, or iPSCs. (The achievement would garner Yamanaka the Nobel Prize in Medicine in 2012.) In this immature state, these new cells are capable of developing into any type of cell in the body, including eggs or sperm. When Yamanaka's method was published, Loring immediately adopted it in her laboratory. It was no longer necessary to get stem cells from embryos; they could essentially grow as many as they wanted from a single skin biopsy. The method was so foolproof that she was soon assigning the job of reprogramming cells into iPSCs to her undergraduate interns. "This, in my field, was like getting a huge present," she said.

The applications for research were seemingly limitless. iPSCs made it possible to generate living mice from mouse skin cells, an achievement that could pave the way for growing replacement organs from a patient's own cells. And because the iPSC lines could be specific to patients, their bodies wouldn't reject them. Loring's lab was soon initiating projects to reprogram the cells of people with Parkinson's into iPSCs, turn these into brain cells, and reintroduce them to the brain for treatment. The possibilities "were the kind of thing you dream about as a scientist," said Loring. "It's like magic."

After feeding giraffes and watching the animals on their trip to the San Diego Zoo, Loring's staff began to consider an unprecedented idea, that iPSCs could be created from the cells of endangered animals being held in the big liquid nitrogen containers within the Frozen Zoo. If this was successful, these new lines of iPSCs could be turned into any cells of the animals, including sperm and embryos that might be used to increase the genetic diversity of existing populations or increase the number of individuals of an endangered species. "If you can make eggs and sperm, you should be able to use assisted reproduction methods. In vitro fertilization [to] create entirely new animals," said Loring. "So we would not just preserve ones that are left but create new combinations of genomes that would add diversity to populations. Diversity is what it's all about."

For Oliver Ryder, the obvious first candidate for such an experiment was the northern white rhinoceros, the rarest and second-largest land mammal on earth. There were only eight individuals alive in captivity and none left in the wild. The Frozen Zoo, however, had tissue samples from twelve different individuals in its collection. Ryder had long been interested in the fate of the subspecies and felt a personal stake in its survival, stemming from his work with Ian Player, the eminent South African conservationist whom he first met in the mid-1980s. Player had been instrumental in saving the southern white rhinoceros, the cousin of the northern subspecies, from extinction. In the early 1900s, there had been only a few dozen or so southern white rhinoceros alive in a remote game reserve called Umfolozi (today it is part of the Hluhluwe-Imfolozi Park). The South African government responded by banning rhino hunting. Player first went to Umfolozi in 1952 as a young gamekeeper and later described seeing his first white rhinos in mythic fashion: "Two bulls loomed out of the mist. I had a perfect view of their physical characteristics as they walked along a ridge. The mouth was square and the nuchal hump between the head and the withers bulged prominently. Flies clung to their flanks and steam rose from their backs. These were truly creatures from a bygone age. The two rhinos grazed as they moved, their heads swinging in a scythe-like motion as they fed on the grass. I watched them move through the gray nthombothi trees into a cluster of candelabra aloes, and disappear into the mist. I had a sudden feeling that my life would in some way be bound up with these prehistoric animals."

By the early 1960s, the animals' numbers had increased to more than 2,000, but the preserve had reached its maximum carrying capacity. There were two options, according to Player: either kill off a number of rhinos or capture them and send them to other places. The first option was unthinkable to him, but the second seemed equally foolhardy. No one knew how to capture and translocate temperamental, 4,000-pound animals. Player pioneered the use of tranquilizer guns to do so and transformed what had been considered a nearly impossible and dangerous endeavor into a relatively easy task. Gamekeepers could now create backup populations in zoos and preserves around the world, ensuring the animals' survival into the future.

One of these institutions was the San Diego Zoo, which received a dozen southern white rhinos in the early 1970s. Soon thereafter, the luck of the northern and southern whites began to change and then reverse. Southerns were rebounding and numbered around 3,000 by 1980, but the northern population was shrinking fast. When they had been discovered in 1900, the northern animals were spread over sub-Saharan Africa, from parts of eastern Democratic Republic of Congo (DRC), to Central African Republic, northern Uganda, Sudan, and the southern edge of Chad, and the animals far outnumbered their cousins to the south. By 1981, these countries had been wracked by decades of civil wars and instability, and only a couple of remnant populations were left. When a comprehensive survey was undertaken in 1983, less than a hundred northern whites were believed to exist in the wild. Player's concern was that people were grouping the two subspecies together and, because the southerns were doing so well, wouldn't support protections for the dwindling northern population.

Player enlisted Ryder in the cause, asking him to analyze the two subspecies' DNA and report on their genetic relationship to each other. What Ryder found was that the two subspecies, separated by a million years of evolution, were about as different from each other genetically as they were from black rhinos. It didn't lead to a different categorization of the species, but it solidified the argument that the northern animals were genetically unique and worthy of protection. In the ensuing years, Ryder began stockpiling northern white tissue samples whenever he could. He traveled to a zoo in the Soviet Union to gather five skin biopsies. Later, the zoo in Khartoum, Sudan, sent the San Diego Zoo a male northern white rhino, adding another specimen to the collection. In all, Ryder gathered samples from a dozen northern whites for the Frozen Zoo. "This is a doomed population, it's widely recognized that they are going to decline further," said Ryder. "Yet there is a gene pool that could rescue the small population in our freezer. Mouse stem cells have been used to produce sperm and eggs. There are several pathways for creating stem cells that could bring back genetic variation that's been lost to the [rhino] population."

There was no funding for Loring's research. It fell outside the parameters of both stem cell research and conservation biology. So she decided to undertake it in parallel with experiments to reprogram human cells for the purpose of regenerative medicine, justifying the expense by arguing that the two things were related. In truth, there was no precedent indicating that what Loring and her staff were proposing was even possible. The method developed by Yamanaka to reprogram cells does so by introducing a set of four genes to a skin cell. These genes, what Loring describes as powerful "master" genes that act as modulators, trigger the cell to transform itself and return to a primitive state. For humans, those genes are taken from embryonic stem cells, but this would be impossible to do when it came to the rhinos. Loring thought reprogramming genes taken from horses might work, but on a whim she decided to introduce the human genes to the rhino cells just to see if they worked. A postdoctoral researcher, Inbar Friedrich Ben-Nun, who'd recently joined Loring's lab, volunteered to lead the experiments using a cell line taken from the youngest living northern white, named Fatu, who was born in 2000 at Dvur Kralove Zoo in the Czech Republic. Several thousand of Fatu's fibroblasts, the cells that make up her connective tissue, were put into petri dishes, and twenty-four hours later Ben-Nun introduced a virus to the dishes. This virus was designed to stick to the surface of the cells and act as a delivery mechanism for the genes, but the initial one failed and a second type had to be tried. Every day the cells in the dishes were provided with fresh nutrients. The genes were not very efficient at reprogramming the rhino cells: out of a million there might have only been ten that transformed. Nevertheless, a couple of weeks later, Ben-Nun started to see dramatic changes in the dishes: whole colonies of shiny, smooth cells were developing. The lab technicians began pulling those colonies out with a pipette, cutting them into little pieces and placing them into new dishes. Within a few months, millions of stem cells belonging to Fatu had grown. Loring sent a few of these back to Ryder's laboratory, where the cells's chromosomes could be checked to ensure they hadn't developed abnormalities. Ryder's laboratory confirmed they appeared just as normal as though the sample had been plucked from Fatu herself.

Loring describes herself as the kind of scientist whose "fortunes are up and down all the time. Either I'm too far out or right in the mainstream." In this instance, she was out in the stratosphere. They had created the first iPSC lines for an endangered species, but there was no clear niche for the publication of research in a peer-reviewed journal. Finally in February 2011, they convinced the editors and reviewers of the journal *Nature Methods* to take a look. When it was published in August 2011, the results were heralded as representing a significant step toward creating regenerative treatments for endangered animals that might be afflicted with genetic diseases or metabolic disorders. And then there was the tantalizing possibility that the cells could actually be urged to develop into rhino sperm and eggs, capable of creating rare embryos. In theory, these embryos could be implanted into a related species, most likely a southern white rhino. The result wouldn't just be cloned rhinos, but all sorts of genetic variations of eggs and sperm from the twelve cell specimens at the San Diego Frozen Zoo. The first step in this process couldn't have come sooner for the northern white rhino population. During the time *Nature Methods* reviewed the paper, the worldwide population decreased by one when Nesari, a thirty-nine-year-old female at the Dvur Kralove Zoo, died of old age.

After the paper was published, however, the rhino iPSCs were frozen at the Scripps laboratory and that's where the experiments stopped. Human patients desperately hoping to have children fund the field of assisted reproduction. This marketplace gives doctors and researchers the incentive to create new methods, but there's no clear economic incentive to making northern white rhinos in the laboratory. Nonetheless Ryder assured me that within the next decade, these rhinos will come out of the freezer. "Reproduction in the current population isn't going to provide a sustainable population," said Ryder. "What we're talking about is the only way to prevent the extinction of the northern white rhinoceros." What no one knows is whether any of the rhinos alive today will survive to meet them. It is a species on the brink of extinction and resurrection at the same moment in time.

* * *

Would the rhinos from a freezer be an entirely new species? Is a rhino born of a laboratory from reprogrammed cells the same as one born of living rhinos? Trying to answer these questions I found myself going down a metaphysical rabbit hole, 2,000 years of debate over the nature of reality, impermanence, and identity that is summed up by a thought experiment called the "Ship of Theseus."

The story goes something like this: Around 350 BCE, the Athenians established a memorial to the naval heroics of their founding king, Theseus. They placed his ship in port, where it stayed for centuries. Over the years, the planks of the ship began to decay, so the Athenians replaced the rotting planks with new wood until eventually all of the wood from the original ship was gone. According to the Greek historian Plutarch, the ship became the focus of a popular riddle among philosophers. Was it the same ship? Had the ship changed, and if so, how? Some philosophers believed the ship was still Theseus's, and others contended that it wasn't the same boat at all. Aristotelians believe the form of a thing is its essence, and according to this logic because the ship had the same exact form as the one that was sailed by Theseus, it was the same ship. Heraclitus, however, might have argued differently. According to him, everything moves on and nothing is at rest; comparing things to the flow of a river, he famously said that no one can step into the same river twice. This sage wisdom seems to prophesy the modern understanding of biology. We know today that all the component parts of our bodies—our cells—are in a state of constant flux, of dying and regenerating. Are we the same person nonetheless? Over generations of human history, natural selection has played upon the DNA in our cells, changing the sequence of proteins in a never-ending rearrangement of molecules that adds up to human evolution. So am I the same species as my distant ancestors?

In 1949, paleontologist Benjamin Burma argued that because species change through time, they can't be the same species from one moment of history to the next, that the whole idea is a construct without reality in nature. This has since been proven to be nonsense, but trying to articulate *why* is difficult. When I tried to think about why a rhinoceros born of stem cells in a freezer would be different from one born in the wild, I intuitively felt the answer had something to do with authenticity. But evolution complicates and obfuscates the idea of any species as something that has an "authentic" identity. This is why scientists and philosophers have come up with over two dozen different species concepts. What persists through time? What *exactly* is a rhinoceros?

Metaphysicists can bring clarity to this question where biologists cannot, but only by grappling with concepts such as space-time dimensions and wormholes. To try to understand these ideas, I turned to French philosopher Julien Delord. In a 2014 essay called "Can We Really Re-create an Extinct Species by Cloning? A Metaphysical Analysis," Delord explained two popular metaphysical stances on species, "real essentialism" and "three-dimensional individualism." The first, real essentialism, is rooted in the ideas of Aristotle and his notion that a horse is a horse because all horses share the same properties and therefore the essence of horse-iness. Some modern philosophers have equated this essence to DNA, but essentialism is mostly considered indefensible after Charles Darwin because the properties of what makes a horse a horse are not static. A genetic trait or mechanism might disappear over generations of horses, but this doesn't mean that the offspring are necessarily a new species of horse. A genetic code might also be shared by different kinds of animals, dogs and wolves for instance, but this doesn't make them the same species.

The second idea Delord wrote about, three-dimensional individualism, is that species aren't really a class of organisms at all, but are better described as individuals. The American philosopher and biologist Michael Ghiselin articulated this idea in the 1960s, and he believed it was not as radical a proposition as one might initially think. Individual organisms like you and me have proper names, and we are made up of parts that are restricted to a particular time and space: I can't be in New York while my feet are in San Francisco. Species also have proper names and are limited by time and space, yet they are considered taxonomic classes, groups of entities that are defined by their shared properties that are true at any time and place in the universe.

In his essay, Delord used these two concepts to try to solve what he calls the "resurrection paradox," the problem of knowing whether a resurrected animal is truly a member of the original species. The conundrum comes down to whether we can transform an evolutionary product (a resurrected rhino) into an evolutionary process (a species of rhino), explained Delord. If you are an essentialist, true resurrection of an animal might actually be possible, particularly if you believe "essence" is equated with an animal's genetic code. A resurrected rhinoceros would have the genome of its ancestors, and therefore be a member of the same species. But if you think of species as individuals, the whole idea of authentic resurrection is impossible. Delord wrote,

> According to this metaphysical stance, when a species goes phyletically extinct (succumbing to terminal extinction), one can make a straightforward analogy with the death of an organism. It ceases to exist both functionally, as there are no more vital relations (reproductive, ecological and so on), and even materially, as no spatio-temporal entity that was part of the species exists anymore. . . . All attempts to resurrect it from a cell or from the genetic information taken from the dead organism is doomed to failure, as this would create a new organism, that is a new spatio-temporally delimited individual, although one very similar in many aspects to the dead organism.

Then Delord introduced wormholes to his analysis and the metaphysical debate over whether reality is best described as having three or four dimensions. In three-dimensional reality, things exist in space but not time. If we think about Theseus's boat from this "endurantism" perspective, wrote

Delord, it's clear that it's not the same boat once even one plank of wood has been replaced, because all its original parts are not present at the same instant in the form of the boat. It's difficult to make sense of species from the three-dimensional perspective because the animals change over generations but, biologists tell us, are still the same species.

Species start making more sense if you think in terms of four-dimensions, what's called the "perdurantist" view, which beautifully solves the paradox of species resurrection, according to Delord. In four-dimensional thinking, entities persist in many possible states by virtue of a kind of temporal wormhole extending in a continuum from past, present, and future. An entity's existence at different stages in time is an aspect of that entity. So even as Theseus's boat is changed plank by plank, it inhabits the same space-time continuum as the original boat—even if all its wood has been replaced over the centuries. A perdurantist might say that a rhino born of a freezer after all other rhinos have died inhabits the same space-time continuum of the species. Its resurrection is as natural as a bear's awakening from the slumber of winter.

All this hand wringing over the metaphysical status of resurrected animals might seem arcane and silly. We don't ask the same questions of human children born of in vitro fertilization or surrogate mothers. But to give humans a moral status is unambiguous. Delord rightfully pointed out that metaphysics of animal re-creation matters a lot when it comes to ethics and how we treat animals that are created by humans. If they aren't allowed full membership to the species they descend from, they might deserve less attention and preservation than "natural" beings. And so whether a re-created rhino is "authentic" or not may matter a great deal in the near future. My head full of these metaphysical debates, I took a plane to Kenya to meet four of the last living northern white rhinos on earth.

✳   ✳   ✳

On a cool and sunny morning in Nairobi, I piled my things into a Land Cruiser and headed north toward Nanyuki, a busy market town at the base of Mount Kenya. In the driver's seat was Kes Hillman Smith, a British-born, Kenyan zoologist, and the world's expert on wild northern white rhinos. We were headed to see Fatu, the rhino whose cells had been used to create the first iPSCs from an endangered animal. Now fourteen years old, Fatu was living on a 90,000-acre wildlife preserve called Ol Pejeta Conservancy near Nanyuki, after being translocated from Dvur Kralove Zoo in the Czech Republic along with three other northern whites including her mother, Najin. At the time, biologists had hoped that their introduction to the grassland of the East African bush would stimulate the animals to successfully mate. But five years later, there still hadn't been a successful conception. Fatu, it seemed, was the last of the subspecies to be born. Kes had been with the rhinos shortly after they arrived at the conservancy, sleeping on a bedroll next to their enclosures. In order to protect them in their crates during the journey from eastern Europe, caretakers had sawed the rhinos' long horns off, giving them a stunted, somewhat sad appearance. Over the next few weeks, Kes and the animals' keepers watched as the rhinos adapted to their new surroundings. Sudan, the only rhino to be born in Africa, was the eldest at thirty-six years old and the dominant male. He seemed to adjust immediately, freely leaving his scent on whatever bush he wanted, marking territory and lying down for a snooze. Suni, the younger male, was different, sniffing around but leaving few scent marks and acting more anxious. Fatu and Najin stayed close together, even though the daughter was well into adulthood. Clearly, the animals had an established dynamic that came from living in close quarters at the zoo. The quickest way to get the rhinos to mate, Kes believed, was to create conditions where social dynamics observed in the wild could develop. Both Sudan and Suni needed to establish separate but parallel territories. That way they could each stake a claim and feel confident mating with the females when they were in estrus; the proximity to another male would spur them on. In the wild, Smith had seen a single rhino population double in the span of eight years, an incredible fact given that females have a fifteen- to sixteen-month gestation period, so she knew what the animals were capable of under good conditions. Clearly something at Ol Pejeta wasn't working.

No one knows more about northern white rhinos in the wild than Kes. In the early 1980s she lived in the eastern corner of the Democratic Republic of Congo (then called Zaire) where the last population was located. She originally planned to stay for a year and ended up staying for twenty-four. It was the only long-standing monitoring project of the animals ever undertaken. Unlike Jane Goodall or Diane Fossey, who both worked in the same corner of Africa as Kes, she never gained notoriety for her work beyond a small, international circle of zoologists and conservationists. This isn't because her work was any less dramatic. During her time at Garamba, a remote and long-neglected preserve, she watched as the rhino population grew to thirty-two individuals, and she then launched a guerrilla war against poachers who hunted and killed one rhino after another. By the time she and her husband left in 2005 there were less than ten rhinos left, and there haven't been any sightings or evidence of them in years.

The pioneering wildlife filmmaker Alan Root described Kes as an unsung hero in one of Africa's fiercest conservation battles, a "pocket-sized Venus, usually dressed in old military gear and boots." She was, Root wrote, like a poster girl for the Israeli Army. I had first heard about Kes's work in Africa from Oliver Ryder, but later I realized that I had actually read about her when I was fourteen years old in *Last Chance to See,* Douglas Adams's funny and moving account of traveling around the world to see endangered species in the early 1990s. The northern white rhino was one of them, and he went to Garamba where he described scanning the horizon with binoculars from a termite hill next to Kes, hoping in vain for a glimpse of a rhino. At that time, there were around thirty rhinos in an area of roughly 650,000 acres.

> Kes is a formidable woman, who looks as if she has just walked off the screen of a slightly naughty adventure movie: lean, fit, strikingly beautiful, and usually dressed in old combat gear that's had a number of its buttons shot off. She decided it wasn't time to be businesslike about the map, which was a fairly rough representation of a fairly rough landscape. She worked out once and for all where the Landrover had to be,

and worked it out with such ruthless determination that the Landrover would hardly dare not to be there, and eventually, of course, after miles of trekking, it was exactly there, hiding behind a bush with a thermos of tea wedged behind the seat.

I noticed there was a thermos in the Land Cruiser with us now, which Kes encouraged me to drink from. In her sixties, she has retained both the petite elegance and simmering intensity of her younger years. Frequently dressed in worn horse-riding breeches and a stylish black leather belt slung low around her hips—the buckle a shiny brass rhinoceros—she drew upon angers and frustrations from decades ago as she told me the story of what went wrong in saving the world's last wild northern white rhinos. "If you think about subsistence living, poaching or hunting is fair enough. People are using traditional methods and hunting for the pot. They need it," said Kes. "The trouble is, it doesn't stop there." We barreled north on a two-lane highway through rich Kikuyu farmland. In the opposite direction was a stream of vehicles returning from an annual charity event farther north in Samburu County. Called the Rhino Charge, the difficult off-road rally race raises money for rhino conservation. The event had raised a record $1.16 million this year, and Kes's son Doungu, named after a river in Garamba, had been a member of the winning car's team.

Rhino conservation in Kenya has entered a period of crisis in the last few years. Poachers killed double the number of rhinos in 2013 than the year before, encroaching on even highly secured preserves with seeming impunity. Kenyan newspapers had begun reporting on the fact that most of these poachers are armed and supported by international criminal networks operating with the help of corrupt politicians. The crime syndicates are feeding a seemingly insatiable hunger in countries like Vietnam and China, where the nouveau riche view rhino horn for medicine or hangover prevention as a status symbol. A kilogram of rhino horn reportedly fetches up to $100,000 in these countries, much more than gold. The number of rhinos being poached is even worse in South Africa; in 2013, poachers were killing on average three rhinos every day.

Kes has seen this all before. It was during the poaching epidemic of rhinos in the 1970s that she had moved to Kenya and was soon hired by the New York Zoological Society and the International Union for Conservation of Nature to do aerial surveys over Africa and analyze elephant and rhino populations. Kes's father was in the British air force and had once given her flying lessons as a birthday present. It was the sort of work she had been preparing and hoping for her whole life. "I suppose I've always loved animals," she said. "I wanted to travel and work in Africa because I wanted to get out of England and be in a more interesting and challenging place." Her doctoral work had focused on electron microscopy in frog hearts, an academic exercise she felt was pointless. "It taught me how to ask questions and formulate investigations, but it didn't seem to have any meaning in the real world. As soon as possible, I got out and into what I considered to be much more meaningful conservation."

What she saw during the aerial surveys was grim. In the Luangwa Valley in Zambia, 2,500 black rhinos had been killed. Three thousand rhinos had disappeared in Tanzania's Selous Game Reserve. The Central African Republic's rhinos were just about gone with about fifty left, and populations in Chad had dwindled toward probable extinction. Southern white rhino populations had rebounded to around 17,000 but now their cousins, the northern whites, were difficult to find. Kes initially estimated that around a thousand were left, mainly in the south of Sudan in the Shambe Game Reserve and at Garamba. Kes became the head of the IUCN's Rhino Specialist Group and proposed that Shambe and Garamba be the focus of funding and efforts to conserve the remaining northern white rhinos. But in April 1981, as civil conflicts were broiling in the region, she flew another aerial survey and couldn't find a single live rhino in Shambe. She began ground surveys in other parts of Sudan, but there were so many poachers armed with military equipment that by 1983 it was hard to penetrate even the national parks. It was becoming clear that the previous population estimate of a thousand northern white rhinos had been optimistic. The total number of rhinos alive, Kes realized, was likely less than a hundred. In 1983, the Second Sudanese Civil War erupted between the Islamic government in the north and the Sudan People's Liberation Army (SPLA). Kes had to give up the idea of working in Shambe and she turned her sight across the border to Garamba in the DRC.

Today, Kes and her husband, Fraser, live in the wealthy Nairobi suburb of Langata, just a stone's throw from where Karen Blixen toiled in the black cotton soil trying to grow coffee, a struggle that would become the basis for her book *Out of Africa*. Their large, well-worn home is concealed from the road by trees. Driving down the lane you pass a barn with horses, and inside the house is a central courtyard overrun by vines, four dogs, and handfuls of cats. In the evenings, their large veranda is often filled with buoyant guests who sit in the canvas safari chairs and watch the wild warthogs and bush babies as the sun sets in the direction of the Ngong Hills. Just beyond the backyard is a 150-acre giraffe conservancy where Fraser rambles with the dogs or Kes rides her horses.

It's an idyllic home but one that seems cast in the shadow of their previous life at Garamba. Entire rooms of the house are filled with papers, research, and funding campaigns for saving the last northern white rhinos. Pictures from their years at the park cover the walls. Kes and Fraser were married on the banks of the Doungu River at Garamba. A daughter came, Chyulu, and then their son. The children both grew up there. In one photograph, Kes grins at the camera as she stands in front of a river, holding a white parasol over her baby girl, who sits in a carrier on her mother's back. Kes looks beautiful in a way that seemed to me to indicate great fulfillment, like she was exactly where she was meant to be. In light of what would come in later years, the photo has a slightly tragic feeling. Fraser intimated this general feeling during an evening stroll through the giraffe conservancy, when he told me we often don't recognize the good times until later, when things are so much worse.

✳   ✳   ✳

Garamba's ecosystem is shaped by rain and fire. During the long wet season from April to November, rain feeds the savanna and the perennial grass, which grows thick and tall until it becomes an undulating, rich green sea. During the dry season, the eight-foot-long grass loses so much moisture in the sun and heat that it becomes combustible and ready to explode. When the wildfires begin, they are ferocious and fast moving, turning the soil gray with the ash left behind. A couple of weeks after the pyrotechnics end, tender green shoots begin to appear, ready for eating by buffalo, elephants, and rhinos, and the whole cycle begins again. Buffaloes were once the most numerous animals in the park. In the 1970s there were around 53,000, in addition to hundreds of Congolese giraffes, kob, warthogs, roan antelope, bushbuck, hartebeest, oribi, duikers, lions, spotted hyenas, and an abundance of hippos and crocodiles. Sometimes the crocodiles in the two rivers running through the park were twenty feet long. The landscape of Garamba wasn't as dramatic as the volcanic mountain regions of the DRC or as impressive as the country's impenetrable interior jungle, but it was unique in its ability to support an incredible density of wildlife. The filmmaker Alan Root wrote that filming the elephants at Garamba was like seeing a remnant of the old Africa, when the continent had "huge herds moving across endless space."

When Garamba had been created in 1938 in what was then the Belgian Congo, the idea was to maintain the natural processes of the ecosystem by completely isolating the land from human activities. The nation's parks, including Virunga, where the country's only population of mountain gorillas live, were to be places in which wild nature could evolve without any interference. When Kes arrived, the brutal irony of this founding vision became apparent. The country's politics, starting with independence from Belgium in 1960, followed by the Simba Rebellion and then an invasion of rebels from Angola, had created instability and a scarcity of funding. Local communities and paramilitary groups used the park's animals as a resource to feed themselves and to create an income. The population of rhinos was as high as 490 in the early 1970s but had since sunk to around twenty. The number of elephants had similarly decreased, from 22,000 in 1976 to 7,000 by 1983. Many of the animals had been pushed into the southern sector of the park by the poachers, who made their camps in the thicker bush toward the north, where they could smoke their meat before taking it to sell in nearby villages or across the border into Sudan.

On the invitation of the government's wildlife agency, Institut Zairois pour la Conservation de la Nature, Kes moved to the park in March 1984 to live full time. Fraser came with her, hired to rehabilitate the park's ailing infrastructure. Fraser was a trained wildlife park manager who had grown up in South Africa and Botswana. He lived in safari shorts and sandals, even when he was darting rhinos or flying a Cessna on poaching reconnaissance missions, and was a natural tinkerer and problem solver, perfectly suited for the challenges of working in Zaire where problems were everywhere. In order to deter the poachers, the Smiths and the park staff began building roads, bridges, airfields, a radio communication system, and strategically located patrol posts. It was an overwhelming and relentless job. In her 2015 book *Garamba: Conservation in Peace and War*, Kes described the difficulty of maintaining a network of roads in the park:

> Under ideal conditions the tractor would cut the grass down the centre of the tracks in June, just as the grass was reaching the height of a Land Rover bonnet and causing the radiators to clog and over heat. If the cut grass was burnt a few days later, the tracks would remain usable through September when another cut was necessary for the centre and sides, to prevent the long grass being bent over the tracks by storms. If all this grass was burnt, the roads would remain open until the dry season when the whole lot would often get burnt regardless of our efforts to control constant burning year after year. It sounds simple, but in practice was seldom possible to achieve. For a start, four cuts down every road added up to thousands of kilometres a year. The tractor was not designed to cope with such a task, nor was the driver, who had to sit day after day in the swirling dust of slashed grass with blood shot eyes peering through thick layers of yellow pollen. Worse still was the task of the aide-chauffeur walking ahead checking for rocks, holes, termite mounds and tree stumps through grass no-less coarse than sugar cane! They stopped only to repair punctures.

Fraser built a mud hut and then a mud-brick house with bookcases and open windows and doors to let in the breeze, and their initial efforts to rehabilitate the park had almost immediate and positive results. Kes was able to do some of the first and only long-term monitoring of northern white rhinos, observing their social dynamics and mating habits across multiple generations. Every month she flew the park's Super Cruiser to conduct a total survey of the population. What she found gave her hope: the population nearly doubled to thirty individuals over the first ten years. With veterinarians Peter Morkel and Billy Karesh, she developed an innovative way to follow their movements using telemetry by drilling into the rhinos' horns and placing radio transmitters inside and up through the center. She was concerned that the small population could result in inbreeding, and began doing biopsy darting and taking small ear notches to send to laboratories in Kenya and Cape Town. Students were coming to the park, including Emmanuel de Merode, who would go on to become the director of Virunga National Park. During this time the poaching was mostly limited to buffalo and the fires from bush camps were easy to detect from planes. Around 1991 the Smiths instituted a rigorous reporting system for the park rangers that gave them ongoing data on the type of illegal activities that they found in the park and the movements of the patrols themselves. "It was a way of being able to analyze and compare information over time and space which was collected in a systematic way," said Kes. "That was incredibly useful. We were getting so much information, which was being fed directly into guiding patrols and improving antipoaching. And we could use it to raise support."

By working with the guards and reinforcing the rules and logistics of antipoaching, explained Kes, virtually all of the hunting stopped. Then unrest came to the park. In 1991, the SPLA took the town of Maridi and 80,000 refugees fled across the border, many of them into the park. Eventually they were resettled nearby, but the ongoing war in South Sudan meant the poaching only increased. When the First Congo War started in 1996, it brought a wave of militias moving through the region; the guards were disarmed and a three-month free-for-all by poachers ensued. The Smiths didn't know it then, but this period of instability would mark a turning point in the fight to protect Garamba, setting off a war between the park rangers and poachers that escalated as the violence around them surged. This was Garamba's curse: it was surrounded on all sides by countries awash in war. The poachers were armed with increasingly heavy ammunitions: automatic weapons, hand grenades, rocket launchers. Antipoaching units, previously eight men and now increased to twenty, were forced to respond by arming themselves more heavily. The fighting escalated. Confrontations with poachers could easily end in casualties on both sides. "They were supposed to give three warnings before they could legally shoot to kill," explained Kes. "But of course they didn't."

Despite the continued aerial surveillance and patrols, the assaults on animals by the poachers were relentless. "We knew it was getting dangerous," said Kes. And they also knew it was only matter of time before the poachers pushed farther south into the park and the long-grass savanna favored by the rhinos. In 1996, patrol units reported hearing gunfire near the Garamba River and discovered two rhino carcasses. The first one was Mai, a young female, freshly killed. Inside her was a male rhino fetus; she had been close to giving birth. The second carcass, a male called Bawesi, had already started to decompose. The rangers cut off Mai's head and brought it back to headquarters. Kes knew some people were poaching animals in the park to feed themselves. But this, she was sure, was the work of the SPLA, which was still fighting a civil war against the Sudanese government in Khartoum, just across the border from Garamba. These men were not poaching because they were starving. "There was a massive amount of food aid that was being flown into Sudan," said Kes. "They were just exploiting their neighbor, selling ivory and rhino horn to get weapons and continue their war." At times her frustration was overwhelming. "You're angry and you think, how are we going to fight this? This is so big," she recalled. "It's a fatalistic anger. So you have to try to create positive force from it, some way to combat it."

The first strategy was to go to the SPLA directly and negotiate to get the poachers to stop. She traveled to the main SPLA headquarters in Nairobi as well as Uganda, where the SPLA had training camps. The army's higher-ups claimed the poaching was just a few stragglers here and there, deserters mostly trying to make an extra buck. They even agreed to do some antipoaching joint operations with Garamba's park rangers. But the Smiths knew things were much more complicated than what the SPLA leaders were telling them. The army had camps just across the boundary of the park, places from which they could cross back and forth with impunity. The park rangers were discovering too many poachers who were Sudanese to believe they were just deserters acting alone. Later that same year, Laurent-Désiré Kabila, the Congolese rebel, began leading his forces over the eastern border of the country from Rwanda. As they marched toward Kinshasa, their presence created a tsunami of military forces fleeing for their lives and pillaging along the way. For the first time, the security situation forced the Smiths and their children to evacuate their home; they hid photos and valuables where they could and flew to Kenya in one of the small planes. After they left, Mobutu's mercenaries took up residence in the park, staying for a few months before Kabila's Alliance des Forces Démocratiques pour la Libération du Congo-Zaïre finally disarmed them and drove them out. In May of 1997, Kabila became president of the newly named Democratic Republic of Congo, and in July the Smiths were able to get back into Garamba. Vehicles, computers, and fuel had all been pillaged. But worse than that, the months of disruption to the antipoaching patrols had led to 6,000 elephants and two-thirds of all the buffalo and hippos in the park being killed. Poaching camps littered the park. Somehow only two rhinos had died and five rhino calves had been born.

They were barely in the midst of rebuilding when the country's second civil war erupted in 1998. President Kabila had tried to force Rwandan Tutsi refugees to leave the DRC and return home. The park's officer in charge of elephants was killed by an army firing squad for his Tutsi ethnicity. It wasn't difficult to understand why the World Wildlife Fund decided to close its projects in the DRC. Luckily for the Smiths, the International Rhino Fund—a conservation organization based in Fort Worth, Texas—took over even as the violence escalated. As Kes explains in her book, three different parallel governments backed by foreign interests effectively controlled the country, and Garamba fell into territory under warlords supported by Ugandan forces. The park was also close to the conflict in Sudan and very near a territory controlled by Rwandan-backed rebels. Dealing with these different factions made it difficult to transfer fuel and supplies and to keep staff paid on time, if at all.

When I asked Kes how many rangers died during this time on anti-poaching patrols, she avoided giving a number. But it was clearly not unusual for rangers to lose their lives or to have to be evacuated after being wounded. It was equally brutal for the poachers themselves. If they were caught alive, they were imprisoned, and the park rangers sometimes tortured them to get information. "We couldn't really stop them doing that, we weren't the authority in that sense," said Kes. "There is a cruel streak," she added, "which seems to have grown in response to the type of colonialism there." During one debriefing, the rangers told Fraser they had killed some poachers in an attack. When Fraser asked how they could be sure the poachers had died, the rangers started bringing in ears cut from the poachers' bodies. Fraser put an end to the practice, but it was a glimpse into how the park itself was susceptible to the unforgiving violence that surrounded it.

For the next few years, even as political instability in the region continued, the population of rhinos held steady against the poachers' incursions. In April 2003, Kes flew one of her regular aerial surveys and found thirty individuals in Garamba. Oddly, it was a cease-fire between the SPLA and the government of Sudan that was the beginning of the end. For years the SPLA had monitored the border between DRC and Sudan, but after the cease-fire they no longer controlled the area, creating a porous boundary that let an entirely new enemy into the park. Smith began seeing herds of elephants gunned down in her aerial surveys; even tuskless females and infants were shot. In early 2004, the rangers got their first sighting of who was responsible: not SPLA but Sudanese men on horses. This was the *janjaweed*, the Muslim horsemen who were creating havoc in Darfur. These skilled fighters could move fast through the park, much quicker than a vehicle or men on foot, and the small planes were useless in cutting off their retreats to the border. They needed a helicopter or two but appeals to the

United Nations—Garamba was in fact a UNESCO World Heritage Site—
went nowhere. In May 2004, an early morning attack against the horsemen in which the park rangers were outmanned and outgunned resulted
in two rangers being killed and others wounded. Three *janjaweed* were also
killed. "It was a disaster," said Kes. "As a result, there was a lot of fear. The
guys were scared and coming up against horsemen who were really brilliant shooters and fighters." By July there were only fourteen rhinos left in
the park, and by December Smith could find only four in addition to four
outside the park's border.

As early as 1995, the government wildlife agency—now called the Institut Congolais pour la Conservation de la Nature (ICCN)—international
donors, and Kes had discussed emergency measures to take some rhinos
out of Garamba if the threat of poaching threatened to wipe them out. It
was an idea Kes endorsed with great hesitation. The rhinos were the flag-
ship species of the park, the reason international interest and donor dollars
were being directed to conserving a small ecosystem in eastern DRC. If the
rhinos were gone, the will to invest in the area might leave with them. She
had long fought for the rhinos to stay where they were in the hopes it would
rally support for the park. But by January 2005 it was clear that if the rhinos
were going to survive they needed to be removed from Garamba. The idea
was to move them to Kenya and bring them back once things were safer.
President Kabila and four joint vice presidents in Kinshasa agreed to the
plan, but the minister of environment refused to endorse it. He insinuated
on live television that the government, ICCN, and the conservation effort
were selling the rhinos to Kenya, where they would be used to increase tour-
ism. The government, wary of its image in the public eye during an election
year, decided to sack the plan.

After this, things became very sinister. A delegation from the government's wildlife organization in Kinshasa was met by machete-wielding mobs
near the park. The Smiths were briefly arrested and accused of sneaking into
the park illegally. The park's warden canceled a donor-strategy meeting at
Garamba. By this stage, the Smiths had already bought their house in Langata, and the donors decided to gather there instead. For many of them, the

politicking by the DRC government and power plays within the ICCN ranks were the last straw. They had stuck with the rehabilitation project through civil wars and unrest, but the latest string of events had spoiled their good-will. The NGOs decided to pull their aid to Garamba and told the ICCN the only way to get it back was for it to sort out its internal problems. By the end of the summer, with no evidence things had changed, the International Rhino Fund—the Smiths' employer—ended their funding permanently. Af-ter twenty-four years, the Garamba project the Smiths had given their lives to was over.

"We spent millions of dollars in the Congo trying to save it in situ," said Susie Ellis, the current executive director of the International Rhino Fund. "And for various reasons outside our manageable interests, it didn't work." In 2008, a team of biologists went into Garamba in search of rhinos. They found nothing. "The thing we all agree on is that the population is extinct. Definitely in the DRC and most likely in Sudan," said Ellis. Like Oliver Ry-der at the Frozen Zoo in San Diego and others, Ellis considers the last rhinos in captivity doomed, and missed opportunities to bring more into captiv-ity from the wild a tragedy. "The view most of us have is it's too late for the northern white rhino. We really messed up. The technologies to bring them back through cloning are long, long away. In the meantime, I think probably the most important thing we have learned is where the mistakes were made. This will never happen again with another species of rhino."

✳   ✳   ✳

We arrived at Ol Pejeta Conservancy in the early evening, just as the sun was starting to cast its last light of the day on Mount Kenya, now looming to our east. The park rangers at the registration gate greeted Kes warmly; she was the lady that only came to see the rhinos. We drove down an empty dirt road through open bush passing herds of buffalo, plains zebras, and deli-cate gazelles munching on green grass. A brisk wind came down from the mountain but the air felt sweet after the congested streets of Nairobi. When we reached the research camp, there were only two other people there, a woman collecting ticks and a young undergraduate student radio-tracking lions. We slept in beds made up with thick down comforters under Masai tartan prints.

By seven the next morning we were back on the dirt road and headed deeper into the 90,000-acre preserve to the area where the rhinos were kept in large enclosures protected by electric fencing and armed guards. Ol Pej, as the locals call it for short, has some of the highest security of any park in Kenya. There is a canine antipoaching unit consisting of German shepherd guard dogs that can detect and give chase to poachers. In 2013 the conservancy crowd-sourced $45,000 to purchase several aerial drones to help patrol the perimeter. But poachers still get in. The previous month they came at night and killed a southern white rhino. "I hate poachers," said Mohammed Doyo, the head keeper of the rhinos at Ol Pej. Doyo has been working with rhinos since 1989, when he was just a teenager and nursed an orphaned animal at the park. He slept in the same room as the baby rhino, which would wake him up in the middle of the night to be fed with a bottle. Doyo has three young children but he said rhinos are his real babies. When the four northern whites arrived at Ol Pejeta, their survival and ability to reproduce became Doyo's personal cause. "When they have a baby, we are going to get Kes and open a bottle of champagne," he said.

As the guards let us into the first enclosure—about 140 acres of scattered brush and grass—Doyo gave Kes an update from the backseat of the Land Cruiser. Four months earlier, the keepers had brought in a young southern white male from a neighboring conservancy and introduced him to Fatu and her mother, Najin, hoping the animals might crossbreed. The male northern whites were proving to be very unreliable studs. Other than a promising instance when Suni, the younger male, mated with Najin, there hadn't been much copulation, and none resulting in conception. It was unclear whether the problem was physiological, environmental, or just bad luck. Had the females developed fertility pathologies? It was possible that years of living in a zoo and not producing high levels of hormones had affected their reproductive organs. At least one aspect of the problem was that Sudan, the oldest and dominant male at forty-one, had an arthritic leg as a result of growing up in captivity. This made mounting the females extremely difficult for him. "Maybe we'll get him a stool," joked Doyo. Meanwhile, Suni, now thirty-six, still appeared to have the finicky and anxious personality of an adolescent.

The decision to try crossbreeding the animals was a drastic measure. It was made by a management committee, including representatives of the Dvur Kralove Zoo, wildlife veterinarian Peter Morkel, who worked with Kes at Garamba, and Martin Mulama, the chief conservation officer at Ol Pejeta. Crossbreeding could speed up the extinction of the species through hybridization, and no one even knew whether the offspring of such a union would themselves be capable of reproducing. But preserving *any* of the species' genes in a living animal was now seen as a better alternative to oblivion. If crossbreeding succeeded, the committee's reasoning went, the offspring might be mated back with the pure northern whites in an attempt to rescue an even larger part of the species gene pool. So in January 2014, the game-keepers moved a male southern white rhino into an enclosure with Fatu and Najin. Soon after, they moved two southern white females into an enclosure with Suni.

We found Suni in the first enclosure, munching on the short grass and lazily shifting from one foot to another. About thirty yards away two female southern white rhinos did the same thing. We stepped out of the Land Cruiser, and Doyo grabbed some hay and waved it around to catch Suni's attention. Because Suni and the other northern whites at Ol Pej were brought up in a zoo, they are used to humans and can be approached under the right circumstances with great caution. Suni lumbered toward us, his head hung low, until he got so close, within a few yards and no signs of stopping that Doyo had to chastise him. "Suni, no!" he shouted. "No, no." The rhinoceros stopped and backed up a few steps to begin munching on his hay, but soon the females were coming over to investigate what all the excitement was about and Suni relinquished his treat to them. "They are very interested in him," explained Doyo. "But sometimes they are bullying him." Doyo tried to interest the girls in other fresh hay he spread nearby, but they seemed intent on keeping Suni away from the food, sending up puffs of dust with their stumpy legs as they chased him about. I hadn't expected that we would be so close to the animals, and trying to dodge all three of them, with sometimes only a small bush in between their hulking bodies and me, was making my heart pound.

Wild rhinoceros have long had a reputation as ill-tempered, dangerous, and stupid creatures. In a periodical about colonial Kenyan history, I read a story about a rhinoceros charging a bus pulled over on the side of the road, thrusting her horn with so much force it went straight through the metal. Sport hunters considered them to have vicious tempers; they would destroy a campfire on a malicious whim. In fact, rhinoceros are fairly docile vegetarians that spend as much as 50 percent of their days snacking on grass and a good part of the rest of their days napping. When I asked Kes what it was like to watch rhinos in the wild, she chuckled and explained that initially it was pretty boring work. Monitoring the rhinos involved a lot of sitting in tall grass being cooked by the sun while the animals comfortably napped under the shade of the only tree. To me, the rhinos seemed to have both sweet and obstinate temperaments. They needed constant chiding. It was like being around overly affectionate dogs that don't know their own strength.

We made our way to the next enclosure, a 700-acre open plain where Fatu and Najin were located. Kes expressed her disappointment at the way things were being run. While all the females were staying in one place, Suni and Sudan were being moved around from one enclosure to the next, making it impossible in her opinion for them to establish themselves in a territory. "It's not reproducing the wild situation in a way that's conducive to mating," she said. The enclosures needed to be parallel but separate from each other, with the males stationary and the females being switched to stimulate their estrus cycles. "*That* could help stimulate them to mate." Doyo encouraged her to appeal to the management committee. The rhinos had been in Kenya for five years and nothing was working, so any new arrangement was worth pursuing. As we stopped the vehicle, Fatu and Najin stood close together about thirty yards away and stared at us before strolling over to see what we were all about. From far away, rhinos look as though they are chipped out of rock. Their shape is part bovine and part dinosaur, and the color of their skin is a monolithic gray. Up close, you can see the skin is textured with ruts and cracks, like the surface of parched earth, and their horns are not perfectly cylindrical but rather like a piece of wood that has been unevenly chiseled to a point. Looking at Fatu, round like a Venus with sleepy eyelids, long black lashes and miniature elephant tail, I thought that rhinos were the most awkwardly beautiful animals I had ever seen. It seemed incredible that evolution had produced a living thing so strange.

The mother and daughter finally decided to meander away; it was almost noon and time for a nap. They both lay down with surprising grace, one arm and leg tucked under their body as though retired on a daybed. We stood a short distance away observing them in silence. Kes seemed like she would have contentedly watched them all day and into the next.

To my untrained eye there were no visible differences between northern and southern white rhinos. I asked Kes about this, and she explained that northern whites have developed distinct characteristics since they split from a common ancestor a million years ago. They hold their heads higher and have shorter head lengths and dorsal concavities, advantages in an ecosystem such as that in Garamba where the grass grows so tall. "These characteristics are worth preserving," said Smith, whether it's in iPSC lines or by crossbreeding. Hopefully, rhinos will one day be returned to Garamba, where the environment can once again select for their unique characteristics. Within a couple of generations, it might be impossible for even a biologist like Kes to tell the difference between a northern white born in the laboratory or one born in the wild.

Whether Garamba, the ecosystem itself, will be waiting for the rhinos to return is a big question mark. A few years after the Smiths left the park, the Lord's Resistance Army, a Ugandan rebel group infamous for using child soldiers and led by the brutal war criminal Joseph Kony, began setting up camps inside Garamba, now managed by an organization called African Parks, and poaching elephants for their ivory. Now, in addition to local and Sudanese poachers, the Ugandan rebels were reportedly using helicopters to wipe out fifty elephants at a time, many of them with a single shot to the head from above. The DRC's army began sending soldiers as reinforcements to the park rangers, but it appears that history is repeating itself. Violence is spilling into the park from north, south, east, and west as paramilitary groups rush to exploit the value of the animals within Garamba's borders. The situation today is arguably as bad as it has ever been, and this is true for much of the wildlife throughout the African continent.

When Kes came to Africa in her twenties to fly airplane surveys, she was part of a wave of passionate conservationists who came to study wild-life, were changed by the experience, and never left. One evening at the Smiths' home, I sat around a dinner table with a few of their friends, in-cluding wildlife filmmaker Alan Root and conservationist Rosemarie Ruf. Ruf, a Swiss national, was traveling through Nairobi from the DRC, where she has lived since 1979, with almost all of that time in the Ituri rainforest trying to protect the elusive and chimeric creatures called okapis. These animals are at least 6 million years old and look like a hybrid of giraffes, horses, and zebras, with the behavior and temperament of shy, docile deer. Illegal gold mining, civil conflict, and poaching within the reserve threaten their existence. In 2002, Ruf's husband, Karl, who directed the okapi preservation project, was killed in an automobile accident in eastern DRC. Then in 2012, six of Ruf's staff and fourteen okapis kept in captivity were massacred by a Mai Mai rebel known as Morgan, a cannibalistic war criminal who liked to call himself Chuck Norris and was an enthusiastic elephant poacher. Ruf happened to be away at the time of the massacre, but she later found out that Morgan was disappointed she wasn't there be-cause he had hoped to publicly rape her. In 2013, the IUCN listed the okapi as endangered. (The DRC government killed Morgan in April 2014.) In Alan Root's memoir, *Ivory, Apes & Peacocks,* he writes about how his life in Africa "has run in parallel with a heartbreaking holocaust as wildlife conservation has proved to be a disastrous failure. The reasons are many, ranging from greed, myopia, and failed policies to the exponential growth of the human population, which continues to sweep away wildlife and wild places." Sitting at dinner, it was impossible for me not to think about the individuals around the table as testaments to the personal sacrifice and tragedy that can befall those who commit themselves to fighting battles, ultimately losing them.

I revisited these thoughts again five months later when I heard that Suni, the young rhino on which so many hopes were placed, died one night in his enclosure at Ol Pejeta from natural causes. Soon after, a male rhino in San Diego, Angalifu, passed away, and the northern white rhinos were down to five. In an e-mail, Kes acknowledged the tragedy of Suni's death but didn't express fatalism over the species' future. She continues to believe that there may be some northern white rhinos left in the DRC, not in the park itself but in the outlying areas surrounding Garamba that are heavily wooded. She was heartened in 2012 by a report of a rhino in Uganda, though it couldn't be substantiated. As optimistic as this might sound, the discovery that some rhinos have survived against all odds wouldn't be unprecedented. There are in fact many instances of this sort of miracle in conservation, when extinct animals reappear long after biologists believed they were gone. In 2000, the Arakan forest turtle thought to be extinct since 1908 was redis-covered in Burma. In 2009, researchers discovered three small-eared shrews of southern Mexico, a creature not seen for 109 years. Recently, an insect called a tree lobster thought to be extinct for eighty years was rediscovered on a remote rock in the South Pacific Ocean. And in 2013 in eastern Borneo, a wild Sumatran rhino, thought to have gone extinct in the area decades ago, was captured on a grainy black-and-white video, a ghost in the for-est oblivious to the wonder of its existence. Some 67 species of mammals have been rediscovered in recent decades. These stories belie the mysteri-ous resilience of some species to persevere against all odds. If you believe in four-dimensional thinking, these miracles make a certain kind of sense. According to perdurantism, the philosophical theory that says things persist through their temporal parts, species could never really disappear and they never go extinct. The space-time wormhole between a species' existence in the past and the present just gets longer and longer. The dead are only ever further away from the moment in time they were alive.

7

REGENESIS OF THE PASSENGER PIGEON

Neo-Ectopistes migratorius

When the Scottish poet Alexander Wilson moved to America in 1794, he fell under the spell of what he called the continent's "feathered tribes." Wilson felt that his love of birds was beyond his control. "While others are immersed in deep schemes of speculation and aggrandizement—in building towns and purchasing plantations, I am entranced in contemplation over the plumage of a lark, or gazing like a despairing lover, on the lineaments of an owl," he wrote. When he was in the wilderness, Wilson described the experience as conversing with the "great Author of the Universe." Each bird contained within itself the mystery of existence, what he called the "incomprehensible First Cause." Wilson traveled thousands of miles by canoe, horse, and foot to find new birds. One year while passing near Shelbyville, Kentucky, he witnessed something astonishing: a flock of passenger pigeons that he estimated must have contained 2 billion birds. The winged cloud of red-breasted fowl took hours to pass over him. Wilson wrote that it was the most incredible thing he had seen since arriving in America.

The size of passenger pigeon flocks was a thing of terror to the early settlers. "There are such prodigious numbers of pigeons that I do not fear exaggerating when I assert that they sometimes darken the sun," said an explorer in the 1750s. Some white settlers believed the sight of the birds was a portent of doom, perhaps a massacre at the hand of American Indians or a coming plague. These flocks produced a volume of noise that was described as unearthly. In 1834, an English geologist traveling in Arkansas described how his horse shook in fear when the birds passed overhead. "When such myriads of timid birds as the wild pigeon are on the wing, often wheeling and performing evolutions almost as complicated as pyrotechnic movements, and creating whirlwinds as they move, they present an image of the most fearful power," he said. "Our horse, Missouri, at such times has been so cowed by them that he would stand still and tremble in his harness, whilst we ourselves were glad when their flight was directed from us."

The only thing more awesome than seeing a passenger pigeon flock in the air was seeing it descend on a forest to nest. The pigeons built as many as fifty nests in each tree and filled them for thousands of acres. The largest nesting on record took place in 1871 in central Wisconsin where flocks congregated over 850 square miles. Oak trees cracked and toppled under the weight of an estimated 136 million pigeons. One hunter described the phenomenon for readers of a local newspaper. After arriving at dawn, there

arose a roar, compared with which all previous noises ever heard, are but lullabies, and which caused more than one of the expectant and excited party to drop their guns, and seek shelter behind and beneath the nearest trees. The sound was condensed terror. Imagine a thousand threshing machines running under full headway, accompanied by as many steamboats groaning off steam, with an equal quota of R.R. Trains passing through covered bridges—imagine these massed into a single flock, and you possibly have a faint conception of the terrific roar following the monstrous black cloud of pigeons as they passed in rapid flight in the gray light of morning, a few feet before our faces.

The nesting of 1871 *was* a portent of doom, though not for humans. As many as 100,000 people traveled to Wisconsin to kill the birds for food and sport. The number of dead and wounded was so great that their bodies littered the ground and the young hatchlings starved in their nests. Some 300,000 dead birds were shipped by barrels east, where they flooded the marketplace and were hawked for pennies. Professional "netters" who had been notified by telegraph of the nesting caught 1.2 million birds. An untold number were shot with guns; one munitions businessman sold three tons of powder and sixteen tons of shot to hunters. "The slaughter was terrible beyond any description," said a witness. By the end of the nineteenth century, any time the pigeons stopped to nest, they were hunted until it became obvious the flocks were not quite so grand anymore.

In 1877, Wisconsin's government officials realized they needed to make an effort to curb the killing, and they passed a law making it illegal to maim, kill, destroy, or disturb pigeons when incubating their young. Some states had moved to protect the bird as early as 1848, but the attitude of most Americans was that something that existed in such great numbers could never be at serious risk. "The passenger pigeon needs no protection," wrote one government committee in Ohio in 1857. "Wonderfully prolific, having the vast forests of the North as its breeding grounds, traveling hundreds of miles in search of food, it is here to-day, and elsewhere to-morrow, and no ordinary destruction can lessen them or be missed from the myriads that are yearly produced." The 1877 prohibition in Wisconsin was barely enforced. Some people predicted a terrible fate for the species. "If the world will endure a century longer, I will wager that the amateur of ornithology will find no pigeons except in select Museums of Natural History," said the writer Benedict Henry Revoil.

By 1899, the last passenger pigeon in Wisconsin was shot. By 1914 there was only one known surviving bird of the species, and she died at the Cincinnati Zoo. R. W. Shufeldt, a taxidermist, conducted the autopsy of the bird, known as "Martha." (Her partner, George, had died a few years earlier.) "In due course, the day will come when practically all the world's avifauna will have become utterly extinct," said Shufeldt after the autopsy. "Such a fate is coming to pass now, with far greater rapidity than most people realize." Shufeldt did something odd and decided not to dissect Martha's heart, to leave intact the vital organ of the last passenger pigeon.

Naturally, just as soon as the passenger pigeon disappeared, the American public underwent a pang of collective regret. They seemed unable to accept the fact that the pigeons, a marvel of the New World, had vanished, let alone that humans were responsible. How could so many birds disappear completely? One theory was that the birds drowned en masse in the Gulf of Mexico. Another was that they had all migrated to South America. Perhaps the pigeons' white eggs, unprotected by camouflage in the wild, caused their extinction. In 1909, the American Ornithologists' Union began offering rewards to anyone who could find a colony of passenger pigeons. Fruitless searches were launched. In 1947, the Wisconsin Society for Ornithology set the record straight, erecting a bronze monument that said, "This species became extinct through the avarice and thoughtlessness of man."

William Beebe, the American naturalist, marine biologist, and bird curator at the New York Zoological Society, recognized that it was not just passenger pigeons that were vulnerable to the thoughtlessness of man. In his 1906 book *The Bird: Its Form and Function,* he wrote that species of birds that survived millions of years of natural selection were now at the mercy of humans, the species that represented the "culminating effort of Nature." "Let us beware of needlessly destroying even one of the lives—so sublimely crowning the ages upon ages of evolving," he wrote. "The beauty and genius of a work of art may be reconceived, though its first material expression be destroyed; a vanished harmony may yet again inspire the composer; but when the last individual of a race of living beings breathes no more, another heaven and another earth must pass before such a one can be again."

Beebe's admonishment would be much quoted in the decades to come, when the number of extinction threats would balloon, one species after another teetering toward oblivion. But he could not have predicted that within a hundred years, scientists would have the technology to bring back the race of living beings that once graced the skies in such great numbers that it made men and their beasts tremble.

✳   ✳   ✳

I met Ben Novak in the fall of 2013 at the University of California, Santa Cruz, just a few days before droves of students would be returning to the campus, where shafts of grainy sunlight pierce the redwood trees. Novak was a visiting researcher at the school's Paleogenomics Lab under the guidance of genomic scientists Beth Shapiro and Ed Green. Shapiro and Green, who also have a family together, use genomics to understand how species and populations evolve through time, and in recent years their work analyzing polar bears, ancient horses, and Neanderthals had gained notoriety. In 2010, they, with the famous Swedish geneticist Svante Pääbo, released the first ever draft sequence of a Neanderthal genome, which they assembled from three females who lived 38,000 years ago in present-day Croatia.

Novak's work at the laboratory was to prepare and extract DNA from passenger pigeon samples taken from some of the 1,500 birds located in museums and collections around the world. The Paleogenomics Lab had samples from sixty-five, some of them over 400 years old, most gathered by Shapiro from the Royal Ontario Museum's collection and a private collection. (Later, the Rochester Museum & Science Center provided specimens that were 4,000 years old.) Novak's analysis of the DNA would provide a glimpse into yet unanswered questions about the passenger pigeon's biology and extinction. When did they become so abundant? How old was the species? Did their evolutionary history reveal a vulnerability to extinction? But this was just an initial stage in Novak's far more ambitious plan. "We will bring back passenger pigeons," he told me as we walked around the laboratory. "When we fail, I will just figure out why and start again. It doesn't matter if it takes forty years or ten years. It will happen." This was before Novak would appear as a central character in a cover story in the *New York Times Magazine* about the Jurassic Park–like possibility of resurrecting extinct animals, but he was already something of an evangelist with an unabashed faith in the idea that de-extinction is imminent, amazing, and a worthy scientific enterprise with great value to conservation.

I arrived on an important day for Novak at the lab. After months of toiling, he would find out whether the passenger pigeon specimens yielded enough DNA to begin piecing together the bird's genome. Thirty samples had been processed overnight and he was waiting to see the results. Once he knew what percentage of each sample was actual passenger pigeon DNA, he would have the best candidate for a genome sequence. The higher the percentage, the faster the process would go. Novak had hopes that several of the specimens taken from bones might deliver high-quality DNA. "No one has ever worked with bones from the bird before; they typically have a lot less DNA and are full of crappy stuff," he said. But the pigeon bones were so old that they predated any population bottleneck created by the large-scale massacre of the birds during American colonization, and might reveal information about the birds' genetics and population structure before they started to disappear en masse. Novak considers this information critical because he rejects the idea that passenger pigeons relied upon large flocks for survival, and that you would need millions of de-extincted birds for them to survive in the wild. There was some research showing that Novak might be right, but no one had the genetic data to prove it.

A few months before I visited him in Santa Cruz, Novak gave a talk at a TEDx event about his plan to bring back the pigeon and was much-quoted in the media. But his status as a spokesman for de-extinction lacked much endorsement from the scientific establishment. Conservation biologists told me that the twitter over de-extincting passenger pigeons and media attention on the project was an "offensive conversation" in light of the real threat of rapid species extinctions in the near future. Even in the Santa Cruz laboratory, Novak was just another researcher in his mid-twenties spending his days extracting DNA—not even very old DNA compared to ancient horses and Neanderthals. When I asked his neighbors in the lab what they thought about his ambition to bring back passenger pigeons, they generously described it as pretty interesting and then added that it was also pretty crazy.

On top of it all, Novak had very few academic credentials, a fact he himself was quick to address. "I have no PhD, I have no publications," he told me. "I'm seen by many as the zealot or overly optimistic dreamer and a lot of times I'm just seen as a joke."

What Novak has is an incredible love, bordering on obsession, for passenger pigeons. "If you love pigeons, it is your life," he said. "I have no idea why. They attract the most emotional, passionate, die-hard kind of people. I'm one of those people, I'm in that group." I heard the same thing during conversations with other passenger pigeon enthusiasts, people who dedicate enormous time and energy to keeping the memory of the species alive and who know Novak, either by reputation or personally. These folks collect specimens and historical memorabilia. Their unofficial orchestrator is Garrie Landry, a Louisiana-born and bred French-speaking Cajun whose website dedicated to the bird's history has become a virtual meeting ground for passenger pigeon lovers. "Let me tell you what I've discovered over the years," Landry explained to me. "I can say with a lot of confidence that many of the people I meet are very similar in that their attraction to the passenger pigeon goes back to childhood. When you're a child, I think it captivates you more—that something in such an incredible number could absolutely vanish." A few years back, Landry attended the first-ever meeting of Project Passenger Pigeon, a consortium that promotes species and habitat conservation through education about the species. After hanging out together, a lot of the participants realized their love of the species goes back to their childhood, when they stepped into the woods and tried to imagine it full of millions of pigeons. Equally captivating was the notion that there had to be *at least* one, maybe even a handful of the birds out there that had escaped extinction. "As kids, they went in search of passenger pigeons because they wanted to prove the adults were wrong," Landry said, laughing. He recalled Stanley Temple, a highly respected ornithologist at the University of Wisconsin, relating his effort as a child to organize the kids in his neighborhood to search for passenger pigeons.

Landry teaches botany at the University of Louisiana in Lafayette, and in his hometown of Franklin, he maintains an aviary of several thousand birds including Gouldian finches, Luzon bleeding-heart doves, Java sparrows and button quails. (He attributes all the mutations of button quails in America to a single import of the bird he made from England in 1991.) But Landry's prized possession is a taxidermy passenger pigeon he purchased on eBay for $3,000. After he bought the bird, he wrote on his website, "Since my childhood days I have always been totally captivated by Passenger Pigeons, now to own one is truly a god send and I am still at a loss for words to describe how I feel." Landry called his bird George and spent dozens of hours tracing the bird's history, eventually figuring that he was probably killed in the wild between 1870 and 1888. Every year, one or two taxidermy pigeons come up for sale. There is no established market price for the stuffed birds; their value is determined by how much desire a potential buyer feels for the specimen itself. Some might go for as much as $10,000. "It is just a pigeon," said Landry. "I don't think there is anything else special about it but for the fact it occurred in astronomical numbers and that we exterminated every single one of them. That's the thing that makes them stand out. There are other pigeons that went extinct but we don't care about those. There are others that are threatened but we're not attracted to those. If that bird had been in small numbers, we wouldn't even think about it today."

At the Paleogenomics Lab, Novak swiveled in his chair and scrolled through e-mails. His iMac was decorated with Transformer toys and next to the keyboard was a bag of homemade molasses cookies, a gift from his grandmother during a recent visit home to North Dakota. He showed me photographs of his paintings, many of them featuring double helixes stretching across the canvas and passenger pigeons perched in the foreground. Novak has floppy brown hair that he sweeps to one side. He wore white running sneakers, jeans, and a T-shirt. He described his upbringing as blue-collar amid a tight-knit, nurturing family; his hometown, Alexander, has a population of 200. Novak's dad was determined that his son would not follow in his footsteps as a car mechanic, and he encouraged an early interest in dinosaurs and science, taking him to museums and subscribing to *National Geographic* magazine. Meanwhile, Novak's grandfather, Anton, a mechanic and tinkerer by trade, passed on his affection for birds. Anton bred chickens, geese, turkeys, guinea fowl, doves, and pigeons in his backyard; at one stage he had 250 breeding pairs of canaries and supplied the regional pet stores.

Alexander's only public school did not have enough kids for sports teams. The cool kids were the smart kids and Novak was one of them. He helped establish what he calls a "freaky dynasty" at the state's science fairs. As an eighth grader, he read an article in *National Geographic* about biodiversity and extinction and came up with a crazy idea: what if the dodo bird could be cloned one day? He prepared an entire research concept around the idea and won best project in his division at the North Dakota State Science Fair. The next year, the science fair was in Minot, North Dakota, and there he found a National Audubon Society book, opening it to a picture of a passenger pigeon. It was the first time he'd seen the bird, and the image gripped his imagination and wouldn't let go. He later wrote that the experience was akin to falling in love. "Like the dodo bird, it was an extinct pigeon. But in shape it looked much more like the doves that flew over my prairie home and the pigeons that strut the streets of every city. But unlike them, it was the most beautiful bird I'd ever seen."

When Novak graduated from his small high school, he left home for Montana State University. He took as many genetics and paleontology courses as he could and became increasingly fascinated by species extinction. When he was twenty-one, he read a book written by an obscure, Ohio-born naturalist named A. W. Schorger, called *The Passenger Pigeon: Its Natural History and Extinction*. Schorger's is the definitive history of the bird, and Novak was deeply moved. "It's one thing to step out into the woods and go, 'Oh my god, this is amazing,'" said Novak. "But what about stepping out into them and knowing that the passenger pigeon once existed there? I step into those woods and rather than having an awe-inspiring experience, I already know that it is a diminished version of itself. It's not as big and grand as it once was."

When it came time for graduate school, he went to study paleogenetics at McMasters University in Hamilton, Ontario. But after a couple of years, Novak began dropping his courses. He felt depressed and increasingly crippled by a sense of anxiety and unhappiness. He had experienced episodes like it before, but this one was inhibiting his ability to study. When he decided to see a psychiatrist, they diagnosed him with an obsessive-compulsive disorder. Listening to Novak talk about his diagnosis, it's clear it was a defining yet liberating moment for him. He began taking medication and started cognitive therapy. And after much soul-searching, he decided he was an atheist. "I decided I wanted to go out and explore and find what truth or philosophy would work. And I wasn't satisfied with anything I came across," said Novak. "And Christianity didn't make sense as an ultimate truth for me." What Novak did believe in was his dream of cloning passenger pigeons. He began hunting for specimens that he might be able to extract some DNA from, scouring the Internet and searching for birds until he found an advertisement for a taxidermy passenger pigeon on sale in Florida and contacted the owner. She had already sold her three birds, but she offered to put him in touch with Garrie Landry, who would most likely know if there were other birds on the market. Landry has met hundreds of bird lovers over the years but no one except Novak had ever mentioned bringing an extinct bird back to life. "It's exciting to think we might be able to bring them back, even if they are relegated to an aviary in a zoo," said Landry. He told his friend Joel Greenberg in Illinois about the young guy who was talking about bringing to life their favorite bird.

At the time, Greenberg, a research associate at the Chicago Academy of Sciences, was writing his history of the passenger pigeon, *A Feathered River Across the Sky,* which represented the first comprehensive contribution to the literature of the bird since Schorger's work. Greenberg offered up a broader history of the bird's place in American culture, pointing out that the pigeon's extinction should be a cautionary tale in an era of accelerated extinction threats. In his opinion, if people knew about this bird and how Americans once dismissed it as too pervasive to disappear, today's crises might have greater resonance for them. "Birding has been the thing that has driven me in life," Greenberg explained to me. "But what is the largest number of birds I've ever seen? Maybe 250,000 sandhill cranes at Platte River in Nebraska. We value what's here but to realize how much more was here before, that this place, North America, had manifestations of life unrivaled anywhere in the world . . ." Greenberg trailed off as his voice warbled with emotion. Finally, he said, "There's no better cautionary tale to the proposition that, no matter how abundant it is, you can lose it." Greenberg and Novak connected over their mutual infatuation with passenger pigeons, despite Greenberg's ambivalence about the idea of de-extinction. Would such a thing minimize the very cautionary tale he was dedicating years of his life to telling? Nonetheless, Greenberg brought Novak on board the Passenger Pigeon Project, a nationwide effort by 160 museums and environmental organizations to memorialize the centenary of the bird's extinction through conservation advocacy and education.

It was around this time that the Bay Area writer Stewart Brand and his wife, entrepreneur Ryan Phelan, were organizing the first gathering of what they named the Revive & Restore project. Brand is known for initiating the campaign to lobby NASA to release the first images of earth from space and has been an outspoken voice in various countercultural movements—from environmentalism to cyber culture—before emerging in recent years as an "ecopragmatist." For Brand, ecopragmatism is about coming to terms with the fact that global warming has forever changed the environment, and it's up to humans to start engineering nature in order to save human civilization. It is not enough to throw a fence around a piece of land for protection. The future, according to Brand, is one in which we need to carefully manage nature or risk losing it.

Brand's roots in the environmental movement are deep. In the 1960s, he studied biology at Stanford University before meeting Ken Kesey and the Merry Pranksters in San Francisco and then launching the Whole Earth Catalog, a guide to tools and products needed for communal or rural living. The catalog was a mirror of the hippie movement's back-to-the-land romanticism. But the format of the catalog was considered radical for the time. It connected ideas, people, products, and consumers around a new environmental idealism; Steve Jobs once described it as the conceptual predecessor to Google. After launching the catalog, Brand hopped from one idea to the next; at one point he wrote a book about space colonies and sat on the board of the Santa Fe Institute, a respected scientific think tank in New

Mexico. In 2010, he took what some considered a heretical departure from his roots with his book *Whole Earth Discipline: Why Dense Cities, Nuclear Power, Transgenic Crops, Restored Wildlands, and Geoengineering Are Necessary.* Critics saw the book as Brand selling out to a status quo, or the dystopian logic that argues large-scale ecosystem engineering is the only way to avert the current environmental crisis. But many scientists, engineers, and ecologists are embracing biosystems engineering or "synthetic biology," a field that integrates engineering science, design, and environmental sciences as a response to climate change. For Brand, it is the philosophical polar opposite of the environmental movement he had helped to define. He and his wife began to explore the notion that de-extinction might be an important strategy in this new geoengineered future, one that could create what they described as "deep ecological enrichment."

Revive & Restore, a nonprofit that seeks to promote conservation genetics, has since co-opted the term "genetic rescue," the same term used to describe the Florida panther conservation efforts of the early 1990s. They have expanded the phrase's meaning from the translocation of populations to offset inbreeding, to a larger spectrum of intervention that includes genomic editing for living species (treating diseases and replacing genetic traits, for example) to full genomic engineering in order to bring back extinct species through cloning, in vitro fertilization, and surrogates. There appear to be few limits to Revive & Restore's ambitions. In addition to convening meetings to advance the field of de-extinction, the project has identified many candidates for de-extinction, a list that now includes marine mammals, plants, insects, amphibians, Pleistocene megafauna, and birds. Among them are great auks, dodos, New Zealand giant moas, ivory-billed woodpeckers, imperial woodpeckers, Carolina parakeets, Hawaiian ʻōʻō birds, Steller's sea cows, Caribbean monk seals, gastric-brooding frogs, Xerces blue butterflies, aurochs, Tasmanian tigers, woolly rhinoceros, and Irish elks. In all of these cases, the organization's rationale is that humans, as the drivers of these extinctions, have the responsibility to deliver environmental justice, to make amends, in other words. "A bit of cloning can get them back," said Brand during a TED Talk in 2013. "Because the fact is, humans have made a huge hole in nature in the last 10,000 years. We have the ability now, and maybe the moral obligation, to repair some of that damage." For de-extinction advocates, our current technological powers mean that the laws of nature are malleable and reversible, and can be bent toward their conception of environmental justice.

On the Revive & Restore website, one of the FAQs is, not surprisingly, about the film *Jurassic Park*. "It was a wonderful movie, which introduced the world to the idea of de-extinction back in 1993. Its science fiction is quite different from current reality, though," says the organization. "First, no dinosaurs—sorry! No recoverable DNA has been found in dinosaur fossils (nor in amber-encased mosquitoes). Robert Lanza observes, 'You can't clone from stone.' Second, the plot of the movie is driven by protecting the commercial secrecy of an island theme park. Real-world de-extinction is being conducted with total transparency. Eventual rewilding of revived species can be no more commercial than the current worldwide protection of endangered species and wildlands. Ecotourism, of course, is a commercial activity often used to help fund the management of protected areas."

These claims to commercial transparency aren't exactly accurate. The same technology used for de-extinction has massive commercial applications beyond conservation. Just a month after Brand gave his TED Talk in 2013, the *MIT Technology Review* published an article revealing how two consultants to the Revive & Restore project—stem cell pioneer Robert Lanza and George Church, a geneticist at Harvard Medical School—were launching a biotech company called the Ark Corporation. The company's key technology would be induced pluripotent stem cells, like those created for the northern white rhino. The Ark Corporation would use them not only to de-extinct animals but also to create genetically desirable farm animals, an enterprise with huge commercial potential. Down the road, that same technology might allow the Ark Corporation to create eggs from a man's skin or sperm from a woman, or let two men and two women have children that share both their genes. Infertility due to age or reproductive issues might virtually disappear. As Antonio Regalado, biomedicine editor of the *MIT Technology Review*, wrote, the de-extinction of animals puts a warm and fuzzy face on scientific testing that has enormous financial potential.

I approached the topic with great skepticism bordering on cynicism. The growing interest in de-extinction seemed to me to echo another initiative, also born in sunny California. In 2013, Google announced the launch of a new medical company called Calico, short for the "California Life Company," whose purpose is to "solve death," much like Brand and others are hoping to solve extinction. This connection between species extinction and human mortality may be more than coincidental. In 1959, Peter Matthiessen drew a parallel between the two in his first nonfiction book, *Wildlife in America*. "The finality of extinction is awesome, and not unrelated to the finality of eternity," he wrote. "Man, striving to imagine what might lie beyond the long light years of stars, beyond the universe, beyond the void, feels lost in space; confronted with the death of species, enacted on earth so many times before he came, and certain to continue when his own breed is gone, he is forced to face another void, and feels alone in time. Species appear and, left behind by a changing earth, they disappear forever, and there is a certain solace in the inexorable."

Brand, Church, and Greenberg would all play a role in bringing Novak to California. At one of the early meetings of Revive & Restore in 2012, Brand and Phelan assembled a small group of scientists, ornithologists, and writers, including Greenberg, Church, and Beth Shapiro at Harvard Medical School. What were the logistics of de-extincting something like the passenger pigeon? Church proposed that it might cost around $1.2 million and five years to build a passenger pigeon genome with technology that he was developing in his laboratory. In *Regenesis: How Synthetic Biology Will Reinvent Nature and Ourselves,* published that same year, Church wrote that the most obvious reason to resurrect extinct species like the passenger pigeon is to "attenuate, even partially, the wave of mass extinction that is currently taking place and is a hallmark of the Holocene—our own epoch. If the continuing loss of countless species is a tragedy, then the introduction of effective countermeasures, and the increase in species diversity that will accompany them, can only be viewed as a benefit." Sure, bringing back species according to our tastes and prejudices will result in anthropomorphized, "boutique" environments, wrote Church, but we *already* reconstruct the world around us to our wants and needs and have done so since the agricultural revolution.

Greenberg left the Cambridge conference and wrote to Novak, "We had this meeting and these people, they are seriously contemplating what you've always wanted to do. You should definitely get in touch with them." Novak was still in Canada where he had finally tested some pigeon samples he got from the Chicago Field Museum, and he reached out to Church, offering his results for research. Church passed the e-mail to Brand and Phelan. A few months later, Novak was working out of the University of California, Santa Cruz laboratory with the financial support of Revive & Restore, taking the initial steps toward sequencing a passenger pigeon genome and bringing back the species, what he hopes to call *Neo-Ectopistes migratorius*.

Novak is well aware of de-extinction's skeptics and has sat in the same rooms with some of them, debating the impact of de-extinction technology. One of these occasions was a conference hosted by Hank Greely, a law and bioethics professor at Stanford University and one of a handful of academics thus far to take an interest in de-extinction. In May 2013, Greely brought together a group of environmental ethicists, conservation biologists, scientists, and lawyers to discuss de-extinction in depth. To Greely's surprise, the Stanford academic community responded negatively to the event. "One famous Stanford biologist sent me an e-mail saying that it is disgraceful that Stanford is even hosting this conference," Greely told me. "There are people who think this is a terrible idea." But Greely is of the opinion that de-extinction is unavoidable. Once the technological capacity is there, the private sector will find a way to capitalize on the cool-factor. There are government officials who see it as unavoidable too. Someone at the California Department of Fish and Wildlife told Greely, "We really need to talk about this before it ends up on my desk." How would resurrected species be categorized under the Endangered Species Act? Could already limited government funds for habitat protection and conservation research be eventually diverted for genomic engineering and cloning?

At the Stanford conference, the participants' concerns focused on whether the ability to resurrect extinct species might make the current extinction crisis seem less calamitous to policy makers and the public. The power of de-extinction could give us an easy way to avoid the tricky political and ethical decisions around protecting species in the wild. People might start putting genetic material on ice for resurrection at a later date as a matter of standard practice, rather than investing in captive breeding programs. Jamie Rappaport Clark, president and CEO of Defenders of Wildlife and former director of the United States Fish and Wildlife Service, told the group there was no doubt in her mind that politicians would take advantage of de-extinction technology to undermine species conservation. "If the definition of 'endangered' is the danger of becoming extinct and you can't prove it because you can de-extinct it, I could see some incredible shenanigans," said Clark, who played an instrumental role in bringing panthers from Texas to Florida in the 1990s. "Revived species are cool and people will pay money to see them," said Clark, "but it won't translate into increased support for wildlife conservation of species that we need to save today." Ronald Sandler, a professor of philosophy at Northeastern University, pointed out that de-extinction does not necessarily preserve the value of the species as a whole, or its relationship to its habitat. And it definitely does not prevent extinction or address its causes.

Jay Odenbaugh, an associate professor and chair of the philosophy department at Lewis & Clark College, raised what I found to be one of the most intriguing ethical issues. Our planet, said Odenbaugh, is increasingly being made artificial through human-driven climate change. He cited a study that said at least 80 percent of the earth's land surface is directly influenced by humans. Why does that number matter? he asked. Because it indicates that humans are controlling the climate and environment on a scale unprecedented in earth's history. If, asked Odenbaugh, we have taken over as creators, those directing the evolution or extinction of life on the planet, how are we to be humble? What stands over us? Humility, suggested Odenbaugh, is important for thinking about environmental values because it is what shows us our place in relationship to life. Without humility, it is too easy to underestimate or overestimate our value in relationship to the universe.

This is far from a new idea—humility has long been an element of the preservationists' argument for caution in altering nature—but at a time when our technological powers have reached heady heights, perhaps this word is at its weakest. To Odenbaugh, the notion that de-extinction was somehow intrinsically bad because it was humans "playing God" isn't as important. It is a "little too late to be worried about playing God," he said. "We've already done it and we're going to have to continue doing it."

Stewart Brand was also at the conference. At the end of the day, a participant asked him whether any of the concerns around the ethics and values informing the emerging field of de-extinction had altered his opinions. "Define moral hazard for me quickly?" he seemed to quip. "It means something right? I looked it up on Wikipedia once." He then offered a defense of de-extinction as a conservation strategy. There are a lot of kids, said Brand, who might be able to see a baby mammoth in a zoo in their lifetime, and that singular experience could bring about a sea change in attitudes toward the natural world. It would give the next generation a chance to adopt a "non-tragic relationship to nature and conservation, with a sense that humans can do it right, even undo serious damage like extinctions that were done in the past."

Sitting in the laboratory in Santa Cruz waiting for the result to come back, I asked Novak what he thought about humanity's increasing management of nature through bioengineering. "Some people don't like how manipulative it starts to sound, that we're going to be creators of what the world is. But we *already* are shaping the planet!" he said. "Now we have to be conscious of it. We should think about how changes we make now will affect not just the next ten, 100, or 1,000 years, but how things we do now will affect the next 10,000 years." The notion that human control over natural processes diminishes nature, through genomic engineering, for example, infuriated him. "This bird genome is 1.1 billion base pairs that were made through the last 4 billion years of evolution of life on this planet. We can sequence it and that's to me *the* most awe-inspiring thing. We get this sequence back and it's bigger than I can look at or read through. It's not smaller than any of us. It's bigger than all of us." All people need in order to see how humbling the power of de-extinction is, according to Novak, is to see a living, breathing passenger pigeon. "This is how we're exploring the universe, of what makes life," said Novak. "It's as big as gazing into the stars."

*   *   *

One day in New York City I set off to find a copy of the book that remains the bible of passenger pigeon enthusiasts. Schorger's *The Passenger Pigeon: Its Natural History and Extinction* is the result of decades of research of the historical record of the species. A trip to the New York Public Library's Science, Industry and Business Library on Thirty-Fourth Street and Madison Avenue proved in vain when the only copy was discovered to have been lost long ago in the stacks. After several phone calls, I finally tracked down a first edition at the Brooklyn Public Library's central branch, where the librarian deemed it so unpopular a reference book she allowed me to take it out on loan. For the next week I pored over the contents while gingerly attempting not to destroy its delicate pages, so brittle they split like communion wafers.

Biochemist Arlie William Schorger, "Bill" among his friends, earned his PhD from the University of Wisconsin in 1916 after completing an obscure thesis on the oils of coniferous trees. He worked for the federal government and various laboratories in the Midwest, publishing a book called, *The Chemistry of Cellulose and Wood* in 1926. Over his lifetime, Schorger held thirty-five patents in the field of wood chemistry, but he maintained a second scientific life as an inexhaustible, obsessive scholar of natural history. The main focus of Schorger's meticulous research was ornithology. He published 172 papers on birds and wildlife and spent years investigating arcane subjects such as the fat content in the feet of gallinaceous birds and making failed attempts to create a cross section of a single hair from the beard of a wild turkey. One of his projects, scanning every newspaper article published in Wisconsin before 1900 for any mention of the state's wildlife, lasted twenty years. When he retired as a chemist in 1951 at the age of sixty-seven, the University of Wisconsin appointed him professor emeritus of wildlife management, a position he held for two decades, donating his token salary to the university's library. Even those who could call themselves Schorger's friends said he was a difficult man to get to know. A political and social conservative, he seldom listened to the radio and didn't own a television; he preferred to spend time with his extensive collection of natural history books or to go bird-watching on Sundays.

Schorger's body of work is notable not only for its meticulous scholarly detail but a near complete lack of descriptive qualities. His first published writing on passenger pigeons was in 1939, in the form of a three-part serial for the inaugural monthly bulletins of the Wisconsin Society of Ornithology. The subject was "The Great Wisconsin Passenger Pigeon Nesting of 1871." In minute detail, he described the killing that took place. "Shipment of 100 barrels per day over a period of 40 working days would give 4,000 barrels, or 1,200,000 pigeons," Schorger wrote. "This figure would be conservative for the total number of killed." But in his later work, there are glimpses of the feelings Schorger might have harbored for a species he likely knew more about than any other person before. Schorger was a child when he first learned about passenger pigeons from an uncle while they were riding down a rural road in northern Ohio. The uncle described how beech forests once grew where farmland now stretched as far as their eyes could see. When the trees shed their nuts each spring, the pigeons came in the tens of thousands to eat the mast. Men stood in the rift of the forest and fired at the passing birds "until the road was dotted with their blue bodies, and more were killed than could be carried." To the young Schorger, the extinction of a species once so numerous seemed incredible, and his sense of awe did not go away with the passing of time. "Deep, youthful impressions are not easily effaced," he wrote.

Schorger's two decades of research about the bird resulted in seven volumes of handwritten notebooks and 2,200 citations in the final draft of the book. There was little doubt in his mind as to what caused the rapid extinction of the species. "The conclusion is inescapable that the passenger pigeon became extinct through such constant persecution that it was unable to raise sufficient young to perpetuate the race," he wrote. Even having written the world's definitive monograph on the species, however, Schorger was not sure whether there was a threshold at which the species could have survived. If a few thousand had been preserved and bred in captivity, would the species have been saved? To Schorger, the question was impossible to answer. He used the case of the heath hen (a species now being studied by Revive & Restore for possible de-extinction) for comparison. Between 1890 and 1916, the population of this North American bird rose from 200 to 2,000 on a preserve. But as a matter of policy, the preserve's managers suppressed any brush fires, which had before then helped create viable habitat for the birds. When an uncontrolled brush fire did erupt in 1916, it ended up killing most of the heath hens. By 1926, the population dipped to fifty and then finally went extinct in 1932. "This species was in danger of extinction for over a century and it is impossible to fix upon a number that was critical," Schorger noted.

One theory that had been around since the nineteenth century, and Schorger addressed, was that the passenger pigeons survival depended on living in the large flocks that had so impressed people like Alexander Wilson. In 1980 a British herpetologist gave the subject new credence in a paper published by *Biological Conservation*. Tim Halliday, based at the Open University in Milton Keynes, England, proposed that the pigeon's decline was too rapid to be accounted for by human activity alone and that the birds were the victims of something called the "Allee Effect," a principle in biology that posits there is relationship between high population size or density and survival in social species. If a critical point of compromise in population size or density is reached, the species can no longer sustain itself. Halliday believed that passenger pigeons were dependent on large flocks for their reproductive success, and once diminished, the birds' breeding rate was insufficient to offset mortality. Soon after the publication of the paper, this idea became part of the birds' lore and was taken up by other biologists. In 1992, a South American biologist, Enrique Bucher, also invoked the Allee Effect when he proposed that the passenger pigeon had a minimal viable population when it came to successfully locating food. Bucher thought that once beech and oak forests in North America were diminished by agriculture and fragmentation in the 1800s, a flock's ability to find food was challenged. As the population declined, a deadly feedback loop was created and the species became increasingly ineffective at this central task. The loss of breeding habitation and "social facilitation at low densities would have been enough to lead the passenger pigeon to extinction even without killing a single bird and despite the existence of considerable remaining forest," he wrote.

Schorger estimated that before widespread hunting by American settlers, the population of passenger pigeons was somewhere between 3 and 5 billion. If people like Halliday and Bucher are correct, and the bird could only survive in such astronomical numbers, the idea of resurrecting passenger pigeons in the twenty-first century begins to look pretty impractical if not downright foolish. Before airplanes, industrial agriculture, skyscrapers, and suburbs, the passenger pigeon was a destructive force in nineteenth-century America. Farmers sometimes sat for days by their newly sown fields armed with guns or sticks to scare away the flocks. The birds were prodigious feeders that could eat almost anything. They preferred forest nuts—beech, acorn, and chestnut, in that order—but they fed on whatever was available: juniper and black gum berries, black cherries, sassafras drupes, sumac seeds, wild grapes, strawberries, grasses, insects, earthworms, and wild rice. The revered naturalist Aldo Leopold called the birds a biological storm, "sucking up the laden fruits of forest and prairie, burning them in a traveling blast of life." It seems impossible that modern Americans, already intolerant of forest fires, floods, and other natural disruptions, would welcome the return of this violent force into their landscape.

This was the first concern Bryan Norton, an environmental ethicist and professor of public policy at the Georgia Institute of Technology, raised when I talked to him about the de-extinction of passenger pigeons. Pigeons may have been useful for the forest ecology, he said, but they were terrible for farmers whose numbers were growing in the nineteenth-century economy. "The cautionary tale is that people weren't so happy with passenger pigeons when they had them. [The birds] totally destroyed farmers' crops. If they had not been eaten like squab, they would have been killed as pests."

Norton believes there is a continuum of appropriate intervention when it comes to genetic engineering and conservation. At one end is the careful use of genetics to avoid interbreeding in endangered populations of animals. In the middle of the continuum is perhaps creatively manipulating artificial traits to help a species survive. "I'm a happy geneticist on that end of the continuum. Further along, I start worrying that what we're doing is virtuoso genetics, not conservation genetics," he said. "We need to really ask, where are we headed with this? Where should our comfort zone be when exploring our abilities to perpetuate species?" Norton is a self-avowed pragmatist within his field, someone who is interested in developing strategies and means for ethics to help solve practical environmental problems. As a result, he is wary of dichotomies, of any argument that depends on black-and-white distinctions of what is natural or unnatural. "You have to do violence to draw a single line and say everything is natural on one side and not on the other," he said. So while Norton is concerned about an era of virtuoso genetics and the economic consequences of passenger pigeons in modern America, he readily admits he would love to see a living passenger pigeon and that such a technological feat could contain valuable lessons for science. "It's a wonderful case to pay attention to," he said. "It's biologically complicated, it's interesting and exciting."

There are colleagues of Norton's who find the enterprise much more odious, even dangerous. To them, bringing back passenger pigeons threatens to subvert the logic that underpins the argument for preserving nature. Erik Katz, a professor at the New Jersey Institute of Technology, is one of them. "If it doesn't matter that we cause the extinction of a species because we can just re-create them in the laboratory," said Katz, "it changes our whole concept of preservation." Katz wrote the earliest dissertation in North America on the topic of environmental ethics, as a graduate student at Boston University in the 1970s. *The Moral Justification for Environmentalism* came out of Katz's interest in medical ethics and how moral theory pertains to individuals who are comatose, senile, or insane—in other words, people who do not seem to possess the requisite status of "persons." This line of inquiry brought Katz to the question of moral obligations toward nonhuman subjects, animals and ecosystems.

Katz was working at a time when the idea of intrinsic value was relatively nascent. Inspired by Holmes Rolston III's early work, he started formulating his own moral theory. To him, an argument that seeks to reconcile environmental ethics with the logic of utility could never hold up, for example, in the context of the Third World, where the need for human survival and environmental preservation are so often at odds. In his search for a nonanthropocentric reason for preserving the environment, Katz arrived at the principle of *autonomy*, a theme throughout his subsequent body of work. Autonomy, says Katz, is the free development of individuals and natural processes—its opposite is domination. When humans begin imposing their ideals and human projects on nature and natural processes, it is a form of anthropocentric domination. "As moral agents, our primary moral goals are to preserve autonomy and to resist all forms of domination, both within the human community and within the natural world," said Katz in his book *Nature as Subject*. Bioengineering to him is another expansion of human power to mold, manipulate, and dominate the environment, as is re-creating extinct species. For Katz, the passenger pigeon project is an enterprise born of human nostalgia rather than a rational conservation effort. "In all of these kinds of cases—whether it's species or ecological restoration or geo-engineering—humans are saying we have the know-how to manipulate the world for our own interests. I don't think that's right," he told me. "These things are ontologically different from the original; they have a different being or essence than the original. The new species is not really a natural species. It may look and act like a species, but it's really a human creation."

✳   ✳   ✳

Ben Novak does not want to create a scientific novelty or a zoo attraction. He wants to repopulate the American skies with flocks of birds that might possibly one day sustain themselves without human intervention. He believes that releasing the species into America's forests will increase the health of the ecosystem, acting much like wildfires do by fertilizing the ground, rejuvenating nutrients, and letting in sunlight for new growth. Biologists have speculated that the biological storm effect described by Aldo Leopold was a sort of ecosystem engineer, causing trees to topple and increasing fuel for wildfires that then set in motion a process of regrowth. The birds' extinction and the resulting dip in acorn consumption might also have increased the number of oak trees, and in turn the populations of white-footed mice and deer, two other significant acorn predators and carriers of the tick that spreads Lyme disease. David Blockstein, a senior scientist with the National Council for Science and the Environment and a passenger pigeon enthusiast, proposed in a 1998 paper for *Science* that the soaring levels of Lyme disease in America are linked to the pigeons' extinction.

Over the course of 2014, Novak and Revive & Restore began emphasizing more strongly the importance of their work for restoring a critical ecological relationship between American forests and passenger pigeons, rather than the more sentimental message of delivering ecological justice for an extinct species. This emphasis on bringing back a whole ecology differentiated them, in Novak's opinion, from many conservation biologists who were too focused on saving a species rather than entire environments; de-extinction was about restoring biological relationships. Novak's work was also beginning to look at the pragmatic aspects of translocating de-extincted birds into the forest, and to this end he was undertaking a risk assessment, methodically reviewing the available data to explore four different scenarios, in a sense building a case for why passenger pigeon resurrection was more than just scientific wizardry. Among the questions he was asking were: What if they just introduced band-tailed pigeons, the passenger pigeon's closest living relative, to the Northeast? What if they did nothing? What if they engineered band-tailed pigeons into passenger pigeons? What if humans could somehow replicate the pigeons' influence on the forest? Novak was also measuring 500 different acorns and chestnuts, with plans to feed some of them to band-tailed pigeons, then planting their poop to try to establish which seeds the pigeons would kill by eating, and which ones might survive to be dispersed and grow into trees.

But how was he *actually* going to re-create the bird? To get a living baby passenger pigeon involves a sequence of technological steps that have never been attempted in the context of de-extinction. Once Novak could see how much DNA was in each of the pigeon specimens, he would pick a couple of samples to sequence. Within these sequences are the clues to what makes a passenger pigeon different from other pigeons, the evolutionary mutations that define it as a species. But this sequence is just a map in a format that cells can't read. In order for the map to be of any use, it would need to be packaged into chromosomes and inserted into a nucleus where it could actually give directions to the cell. No one knows how to do that. Novak's strategy would actually bypass this problem. Instead, the DNA of band-tailed pigeons would be edited to include critical passenger pigeon mutations. This would be done using the genomic editing tool known as CRISPR-Cas9, which was pioneered by molecular scientist Jennifer Doudna and her colleagues at the University of California, Berkeley. A naturally oc-curring protein in some bacteria, CRISPR-Cas9 has the ability to scan its host genome for foreign DNA and then "cut" that DNA out of the genome. Described by some as "molecular scissors," scientists now know how to di-rect CRISPR-Cas9 to any desired location within a genome—plant, animal, human—to cut or modify particular genes. At Harvard's Wyss Institute, George Church is experimenting with CRISPR-Cas9 technology to engineer mosquitoes so that they can no longer transmit malaria, and control gene expression in human cells that inhibits the effectiveness of the HIV virus. Another project of Church's is using the tool to engineer elephant cells with the goal of recreating mammoths.

In Novak's case, he would use CRISPR-Cas9 to create a passenger pi-geon by editing band-tailed pigeon DNA in primordial germ cells—the kind that develop into sperm and oocytes—so that the cells contain passenger pigeon traits such as longer tail feathers or red breasts. Those living, chimerical cells would then be inserted into an embryo. It's theoretically possible that passenger pigeons might be hatched from band-tailed pigeon hosts. In 2012, researchers at a California-based commercial company called Crystal Biosciences announced they had successfully created a way to propagate endangered birds by using surrogate birds of a different species. The researchers created "interspecific chimeras" of male guinea fowl, whose testicles had been colonized with germ cells from a chicken. These chicken cells became functional "chicken" sperm, which was then inseminated into a chicken embryo to produce normal chicken offspring. The company estimated that thousands of endangered birds could be produced in a short time using just a hundred host birds, and the company is now partnered with Revive & Restore with the goal of hatching passenger pigeons using this technological blueprint.

In Novak's opinion, once he has created a living pigeon, it would take just a few years to condition an entire flock of pigeons capable of sustaining themselves. His plan is to introduce baby pigeons to rock pigeons that have been cosmetically "painted" to resemble the extinct species. After a couple of weeks, these surrogate parents would be removed and the young birds would band together as a flock of juveniles, just as Schorger described happened in the wild. Then, using homing pigeons that had also been cosmetically painted, adult flocks of "passenger pigeons" would be sent out to absorb the juvenile flocks, showing them the way to aviaries loaded with food in the forest. "After two or three years, we slowly take away surrogate flocks, take down the aviaries," said Novak during his TEDx talk. "We get to witness the passenger pigeon rediscover itself in the New England and Great Lake forests of North America." After that talk, a handful of people started contacting Novak, offering to volunteer in training passenger pigeons or releasing them onto their land. The issue of private versus public land would be a critical issue for the project. "No one is going to care if they devastate a few square miles of state or national forest," said Novak. "But the moment it's private land, it's going to be different."

✳   ✳   ✳

When Novak got the results of the data processing back, nearly every tissue specimen was higher in quality than he had hoped. Ten of the specimens showed over 60 percent pigeon DNA. The best one—likely the candidate for a whole genome sequence—was from the Royal Ontario Museum. It was a female bird, shot by a man named William S. W. Grainger along the shore of the Don River in Toronto in 1871—the same year the largest pigeon roost in history was recorded in Wisconsin. It happened to be the first tissue sample Novak had processed in the lab and therefore had the name BN1-1 for "Ben Novak's Extraction 1, Specimen 1."

What Novak still didn't know was this: How many passenger pigeons are needed for the species to survive? Over the next year, his efforts focused on creating the data to answer this question. While some biologists believe the species needs high numbers in order to survive, others have found evidence disproving the idea that passenger pigeon numbers were always in the billions. In his 2006 book *1491: New Revelations of the Americas Before Columbus,* journalist Charles Mann tells the story of how two scientists traveled to Illinois in 2003 to scour the ruins of Cahokia, one of the largest ancient cities in the Americas. They were looking for passenger pigeon remains, but what they discovered instead was surprising. Rather than thousands of pigeon bone remnants in the ground, there were few traces of the birds to be found. How could this be? If the passenger pigeon was as ubiquitous as American settlers said they were, why was there so little evidence that the birds were a significant food source for Native Americans?

Archeologist Thomas Neumann had wondered the same thing back in 1985. His analysis of prehistoric sites in the eastern United States led him to believe that passenger pigeons had a small population size relative to their later abundance. Neumann concluded that prior to European contact, passenger pigeon numbers were kept in check through what he described as human-wildlife competition. Native Americans kept pigeon numbers low because they consumed the same mast as the birds did: beech, acorn, and chestnut. This ecological relationship was disturbed after first contact with Europeans led to disease outbreaks and mass death among the indigenous tribes. The billions of pigeons settlers witnessed were actually an "outbreak population," wrote Mann in a *New York Times* op-ed, the result of increased food supply rather than an expression of the species innate biological evolution. And the same was true of bison, elk, and moose. "The huge herds and flocks seen by Europeans were evidence not of American bounty but of Indian absence."

Novak believes that you don't need billions of pigeons to impact the forest, but that density might have something to do with their ability to act as ecosystem engineers. By early 2015 he had two full band-tailed pigeon genomes sequenced as well as two passenger pigeons. The results had completely changed his understanding of the pigeons' evolutionary history. "We now know how old the evolutionary lineage of the passenger pigeon actually is," Novak told me, pausing with excitement. "The band-tailed pigeon and passenger pigeon split 22 million years ago." This is truly an incredible number. It's not that all other birds are younger; hummingbirds, for instance, are around 42 million years old. But there are 338 species of hummingbirds that have evolved since then, and there has only ever been one species of passenger pigeon. What Novak and his advisers at the Paleogenomics Lab believe is that around 25 million years ago, there was a giant population of pigeons in the forests of North America. Then around 22 million years ago the Sierra and Cascade Mountains began to form, splitting the population in half. While the ancestral band-tailed pigeon spread to South America and the Caribbean, eventually evolving into seventeen different subspecies, the passenger pigeons survived as one species for 22 million years. As Novak pointed out, every other genus in the pigeon group has split into subspecies. "The *Columba* lineage might be around 30 million years old, but it has like forty species in it." Novak's voice rose with enthusiasm. "Even the mourning dove genus has five or ten species!"

Throughout their history, the passenger pigeon population had remained remarkably stable. There was no indication that ice ages or climate variability had created genetic bottlenecks. This had Novak scratching his head for a while. If half of the pigeon's native range was covered in ice tens of thousands of years ago, why hadn't it resulted in a population dip? Novak suspected that the birds' ability to eat almost any type of food—not just mast—was the answer. "Through all these forest changes, the birds were unaffected. We're trying to figure out what that means and it just supports this notion that the bird was kind of a superspecies," said Novak. "The pigeons didn't have a migratory pattern, they were nomadic. So if the forest changes, they just moved and changed their diet. In the 1800s they were observed to eat acorn and beech. Twenty thousand years ago, it would have been pine and spruce. They were supergeneralists that are capable of responding to environmental change. All that matters to pigeons is the amount of forest that is available to them directly, and the forest's productivity."

Novak was still a few months from being able to fully compare band-tailed and passenger pigeon genomes to begin identifying what exactly made the species different from each other. There seemed to be about a 3 percent difference in their DNA, and it was increasingly likely that significant differences were social: how densely they roosted and nested together. Whereas band-tailed pigeons spread out, passenger pigeons were known for assembling in the thousands in just a few trees. Novak hoped the genome editing experiments might begin sometime in 2016, but this wasn't about making band-tailed pigeons look like passenger pigeons anymore. He would need to understand whether certain genetic differences were linked to different social behaviors in passenger pigeons and whether introducing mutations to the genome would be enough to coax these behaviors out of the new birds. It would be an enormous challenge. It was in a sense attempting to replicate 22 million years of evolution in a laboratory. Perhaps the birds' social behavior was more nurture than nature; he just didn't know. Nonetheless, Novak is still confident that within a decade he will look into the scarlet eyes of a simultaneously new and old feathered tribe.

To me, the incredible thing about the 22 million years of evolution, and the pigeon's amazing resilience over that time, was that it meant that after thriving for eons, they had disappeared in one-thousandth of 1 percent of their total evolutionary history. It is now almost surely true that human predation caused their demise. For million of years, forests in the Northeast had hosted the birds, and though they have only been gone for a hundred years, we have almost completely forgotten that they were here.

Before I left the laboratory in Santa Cruz, I asked Novak what he thought Schorger would think of his endeavor. Novak paused. "I'd like to think that he would really support it." He told me about an e-mail he received following his TEDx talk, from a man named Ed Lyle in Florida. Would Novak like a signed, first edition copy of Schorger's venerated book on passenger pigeons? Lyle, it turned out, had been Schorger's personal caretaker as a young biology student at the University of Wisconsin, after the natural historian began losing his sight and hearing. In the summer of 1971, just a year before he passed away, Schorger signed his personal copy and gifted it to Lyle. Now Lyle wanted Novak to have it. "My price for the book is your continued passion and commitment for what you do," he wrote. "This book belongs more to you than to me." Lyle later told me that he believed Schorger would embrace Novak's work because he viewed the extinction of the pigeon as an unnecessary and unfortunate loss, and that he would welcome the opportunity to right a wrong.

8

NICE TO MEET YOU, NEANDERTHAL

Homo neanderthalensis

O f all the possibilities of de-extinction, perhaps none is as ridicu-lous, alluring, and potently symbolic as the possibility we could one day resurrect our closest relatives, the Neanderthals. Our in-terest in Neanderthals began the moment the first fossilized re-mains were found in Germany in 1856. What were they like? Why did they disappear? By the early twentieth century, fiction featuring Neanderthals revealed an enduring fascination with our extinct cousins and their poten-tial resurrection. In 1939, L. Sprague de Camp, an aeronautical engineer and science fiction writer, wrote "The Gnarly Man," a short story about Clar-ence Aloysius Gaffney, who works at a freak show in Coney Island but is actually a 50,000-year-old Neanderthal who became immortal after being struck by lightning during a bison hunt. "What did happen to your people?" asks the woman who discovers Gaffney's secret. "The tall ones were pretty crude, but they were so far ahead of us that our things and our customs seemed silly," Gaffney tells her. "Finally we just sat around and lived on the scraps we could beg from the tall ones' camps. You might say we died of an inferiority complex." In 1958, Isaac Asimov published *The Ugly Little Boy*, a story of a four-year-old Neanderthal who is kidnapped and brought into the twenty-first century. The 1984 film *Iceman* is about a Neanderthal who is resurrected in modern times when he is unfrozen from a block of Arctic ice. For decades now, the prospect of meeting our evolutionary cousins has been as tantalizing a prospect as the existence of extraterrestrials or time travel.

Most recently, this enduring idea of Neanderthal resurrection was given voice by none other than Harvard scientist George Church. During a 2013 interview with the German magazine *Der Spiegel*, Church talked about the possibility of re-creating Neanderthals and how introducing their genetic diversity to the modern world might represent a strategy for avoiding societal risk. Church was speaking to the fact that, counterintuitively—considering our population size—modern humans have much less genetic diversity than many species, including chimpanzees and penguins. "They could maybe even create a new neo-Neanderthal culture and become a political force," said Church. In his book *Regenesis*, he wrote, "the question arises whether we have an obligation to bring these creatures back, not as circus sideshow attractions but as a part of a focused scientific attempt to increase genetic diversity by reintroducing their extinct genomes into the global gene pool."

The technology for Neanderthal resurrection is almost identical to the plan to bring back passenger pigeons. Start with the physical genome that most closely resembles the Neanderthals (*Homo sapiens sapiens*) and manipulate the Neanderthal genome into existence. Neanderthals were among the first extinct species to have been studied by geneticists; in 1997, researchers recovered mitochondrial DNA from a Neanderthal specimen, and Svante Pääbo, the well-known evolutionary geneticist at the Max Planck Institute for Evolutionary Anthropology, has since spent years mapping the Neanderthal genome. But actually bringing a Neanderthal embryo into the world would require an incubator, and this is where Church unfortunately put his foot in his mouth, suggesting that there would be no better candidate than a modern woman. After the interview was published, the media went crazy. "Wanted: 'Adventurous woman' to give birth to Neanderthal man—Harvard professor seeks mother for cloned cave baby," splashed Britain's *Daily Mail* newspaper. Later (as his own book was being published) Pääbo blasted both the technical feasibility of Neanderthal cloning and its moral implications in a *New York Times* op-ed. "Neanderthals were sentient human beings, after all. In a civilized society, we would never create a human being in order to satisfy scientific curiosity," wrote Pääbo. "From an ethical perspective it must be condemned."

Church had in fact publicly protested after the *Der Spiegel* interview that he doesn't advocate resurrecting Neanderthals, and he did so again when I spoke to him. "I'm not bringing back the Neanderthal," he said. "I don't think we have a strong argument yet. We're not even ready for [human] cloning for other reasons, though we are ready for human modification—genetically modified humans are in existence now." (Church was referring to a 1997 experiment in which at least seventeen American babies were born from embryos manipulated to contain genetic material from three parents.) Church clarified his position: he believes we need to talk about Neanderthal cloning in order to make a decision about whether or not to do it in the future. Church's understanding of genetic technology and its capabilities are arguably light-years ahead of the general public; for him, the era of faster, cheaper genomic engineering is already upon us, and the future of Neanderthal resurrection can no longer be considered the realm of science fiction. "One of the disciplines in my laboratory exercises is we try to anticipate things that can go wrong in new technologies rather than casually dismiss them," Church said. Fundamentally, he believes that our ability to use technology to engineer nature *is* itself natural. "An ant can make an ant hole, that's natural, and if we make a skyscraper, that's natural. But it's not ancient," he told me. "I think overall the exploration of our planet and other planets, and exploration of nature and the changing of it—small changes or big changes—help us to appreciate just how vast the whole thing is. It becomes more complex and more diverse the more we change it."

When I began to scratch below the surface of the sensational aspects of Neanderthal resurrection, what I found were questions uncannily similar to those we face as we consider de-extincting other species. Is this a form of ecological justice? Or just the ultimate bid for our domination over natural laws and, symbolically, our own mortality? To begin to answer some of these questions, we have to go back some 24,000 years to the moment of Neanderthal extinction.

Archeologists in the nineteenth century believed that Neanderthal fossils were our human ancestors that had evolved and assimilated into our gene pool. But genetic analysis tells us that Neanderthals were a distinct evolutionary lineage that, rather than adapt or mutate into a subsequent species, underwent a true extinction like the dinosaurs or woolly mammoths did. At one time, the Neanderthals ranged as far east as Siberia all the way through northern Europe and the Mediterranean. Their existence likely centered around small territorial homelands, generally no more than forty-five square miles where they hunted large animals such as woolly rhinoceros, mammoth, bison, horse, reindeer, and wild boar using wooden spears outfitted with stone spear heads. Archeologists have some evidence from dental wear patterns that there were gender divisions among Neanderthals when it came to food and survival, but women and children most likely still participated in hunts and harvested the meat. For the most part, Neanderthals were a species that demonstrated few innovations over tens of thousands of years; the archeological record indicates that the last Neanderthals lived somewhat similarly to the first Neanderthals, relying on large animals for the majority of their food, inhabiting somewhat small territories with insular groups, and relying on stone technology and fire for survival. But they survived for a very, very long time, more than 300,000 years. This fact is confounding. If Neanderthals were so stunted culturally and technologically, how did their species survive that long?

In recent years there has been a flurry of discovery and rethinking about Neanderthal existence that challenges the notion that they were a species without language or significant intelligence. Some of the most exciting and critical aspects of this new research centers around the countless stone tools Neanderthals left behind that have survived the eons. These tools, it turns out, are complex and challenging to replicate even for a computer-wielding genius species like *Homo sapiens sapiens.* There are basically two techniques in Neanderthal stone technology, one simpler and the other so advanced it can be accurately described as stone engineering. In the first and oldest technique, Neanderthals chipped small flakes from a single stone to sculpt a spear point or hand axe. Then in the Middle Paleolithic, Neanderthal stone knapping took a considerable leap: the flakes themselves became the goal rather than the byproduct of the sculpture.

There are a handful of people in the world who know how to make this second type of tool, called Levallois cores after the place in France where they were first found. One of them is a young American anthropologist, Metin Eren. Over Skype one morning from his office in Coventry, England, Eren held up a dark rock the size and shape of a mango to his computer camera. The surface of the rock was roughly textured all over; he had clearly chipped away small pieces off the surface using another stone. "Middle Paleolithic Neanderthals removed flakes in such a way that they created a gentle convexity on one side of the rock," explained Eren. Once the rock takes on this convex shape, the stone knapper can strike the end of the rock with significant force, a single blow that causes the bottom of the rock to fall away into a large flake with a sharp edge all around. The flake is symmetrical, reducing torque and allowing more force and efficiency when it is used to cut. "The flakes have design properties that we don't see earlier on. You can resharpen these edges many times, and they are ergonomic because the center of mass is right in the center," said Eren. "They were engineering their tools."

In their book *How to Think Like a Neanderthal,* anthropologist Thomas Wynn and psychologist Frederick Coolidge, both at the University of Colorado at Colorado Springs, believe the *act* of stone knapping itself can tell us a lot about the Neanderthal mind and its abilities. The process requires significant motor memory and most likely thousands of hours and tens of thousands of repetitions to achieve efficiency. There is an overarching hierarchy to the task, with a set goal and techniques to achieve that goal. The Neanderthal capacity for mastering this technical thinking was likely no different from modern technical thinking, such as a blacksmith or woodworker utilizes in their work. Eren himself, an accomplished piano and soccer player, said the process of learning stone knapping was similar to becoming good at an instrument or sport. "I would practice really hard at it and then take a couple days to recover, to let my mind figure out what was going on, and my muscles recover," said Eren. It took him roughly eighteen months to master the skill.

Levallois cores required advanced planning, technical skill, and most likely a form of language and speech in order for Neanderthals to pass on the information required to replicate them. Not only were they engineering their stones, Neanderthals were attaching spears and arrowheads using glues that required several stages of production and the use of fire as a transformative agent. These discoveries have changed the way we think about Neanderthals, said Eren. It's no longer possible to argue they were an unintelligent, maladapted species; they clearly utilized technology in ingenious ways. There is now so much new research unveiling this intellect at work that it has led some researchers to assume that the species was just like us. "We don't really know that either. The truth is probably somewhere in the middle," cautioned Eren. "But from an evolutionary perspective, a lack of change in Neanderthal technology suggests extremely good adaptation. Rather than thinking of them as stagnant or nonadaptive, the fact we don't see change over time indicates we should think of them as well adapted to their environment."

The logical next question is whether the Neanderthals' cognitive abilities led to symbolic thinking or the creation of religious and narrative traditions. We have long presumed that our own species is a pinnacle of evolutionary complexity, with the distinctive capacity for language and symbolic thinking separating us from other species. These are the qualities that we assume *make us* human. But what if Neanderthals had some of these abilities before us? There isn't any evidence of Neanderthal cave paintings or artifacts, but this could reflect the fragility of the archeological record more than the absence of Neanderthal culture. "They lived in Ice Age Europe where things were a lot damper and wetter than in Africa; soils are different," said Eren. "All of those things can have an effect on what is preserved and what isn't." What *has* been found points to some interesting possibilities. Archeologists have uncovered "crayons," sticks of pigment that Neanderthals may have used to color their skin, and seashells that appear to have been painted and worn as ornamentation. At a site in Sima de las Palomas in Spain, archeologists found two panther paws alongside a half-dozen Neanderthal remains, a possible indication that the bodies were buried ritualistically.

One of the greatest challenges to the belief that Neanderthals lacked language is Svante Pääbo's genome mapping project. Along with Ed Green at the Paleogenomics Lab in Santa Cruz and other researchers, Pääbo discovered that Neanderthals had the FOXP2 gene, known to be directly involved with speech and language and previously believed to be unique to humans. This doesn't tell us much about how developed the Neanderthal capacity was for language, but it indicates an ability. Then Pääbo discovered another clue in his genomic analysis: modern-day humans of European or Asian descent contain pieces of the Neanderthal genome in their DNA. This means that at some point in our evolutionary past, Neanderthals and humans interbred with each other.

Pääbo himself had been dubious of the possibility of interbreeding; there was no evidence for it. But after comparing the results to the genomes of five modern humans, his team discovered that Asians and Europeans share around 1 to 4 percent of their DNA with Neanderthals. Now, anyone can pay $100 and send a swab of their cheek cells to the California company 23andMe for analysis to find out what percentage of their DNA is Neanderthal. (I did it and have 2.4 percent.) The discovery that modern humans had once interbred with what have long been considered a less-evolved species upturned long-held assumptions about our ancient history. At least a hundred thousand years ago, probably somewhere in the Middle East during a period when our human ancestors and Neanderthals had similar technology and our populations overlapped, we interbred and our offspring were fertile and flourished. This extraordinary convergence in our evolutionary past opens the door to the possibility that there was not only communication between Neanderthals and our ancestors, but also a cognitive likeness.

The brilliant scientist and writer Stephen Jay Gould bemoaned the entrenched belief that evolution is a march of progress—a stooped, hairy figure of a primate growing into an ape-ish Neanderthal and eventually an upright, civilized, modern human. This concept of historical progression is ingrained upon us in elementary school, and it makes evolution seem intuitive and completely rational: human consciousness is so complex in its powers that it must be the apex of an inevitable trajectory. But Gould believed that the paleontological record reveals this concept of evolution is a sham. Life on earth, insisted Gould, is *not* shaped like a cone that extends upward toward increasing diversity and complexity, eventually resulting in us. Gould's proof was the Burgess Shale, a fossil field in British Columbia first discovered in 1909. Over 530 million years ago, tens of thousands of marine creatures were fossilized in a limestone quarry that preserved not only their bones but also all of their soft tissues in exquisite detail. These creatures evolved during the Cambrian explosion, when life on earth underwent massive diversification and the blueprint for virtually all modern animals appeared. Initially, paleontologists thought these fossils belonged in existing taxonomic categories, but then in the 1970s, they discovered this was a colossal mistake. The Burgess Shale contained a range of original anatomical designs unparalleled today, indeed, a greater diversity than all the marine life put together in our modern oceans. These fossils are proof, as Gould described, that evolution is not a narrow beginning and a "constantly expanding upward range." Rather "multicellular life reaches its maximal scope at the start, while later decimation leaves only a few designs." If you replayed the tape of life, Gould argued, the chances of humans evolving in this lottery of decimation were slim to none.

The old story of Neanderthal extinction is that our ancestors' greater cognitive ability and technologies gave us an evolutionary advantage to survive climate changes, and we spread across the globe. But as the Levallois cores indicate, Neanderthal stone knapping was also a sophisticated technology that gave them tools well adapted to their landscape. "It's very difficult to say that Neanderthal technology did or didn't help or hinder their extinction," said Eren. "Archeologists who make those claims are going out on a limb." So why *did* Neanderthals go extinct?

Some scholars believe that the last stand of the Neanderthals took place in present-day Spain or Portugal on the Iberian peninsula. Southern territories like these might have acted as refuges during periods of extreme cold and glaciers for tens of thousands of years, and by the Late Pleistocene, the climate was beginning to cool again. Animal abundance might have shifted or even decreased, and now Cro-Magnons, our ancestors, arrived in Europe and were probably hunting in the same territories. Cro-Magnons used spear-throwers and fishhooks and probably expanded quickly; for every one Neanderthal in the prehistoric European landscape, there may have been ten Cro-Magnons. "In direct competition with Cro-Magnons [Neandertals] needed to shift focus to new foods and invent new techniques," write Thomas Wynn and Frederick Coolidge. "Neither ability was a well-developed component of Neandertal cognition. Or their inability to respond may have had nothing to do with cognition. Their territorial communities were small and scattered, and when a territorial community became isolated from others, with no access to mates, it soon perished. After a few thousand years of this, the Neandertal population was reduced beyond the point of possible demographic recovery." My own mitochondrial DNA reveals that my matrilineal ancestors may have been a part of this history. Some researchers think that my genetic group, U5b2a2, was one of the hunter-gatherer peoples that took refuge in southern Europe during the same ice age that finished the Neanderthals. When the ice began to retreat 15,000 years ago, these Mesolithic people may have been among the first to repopulate the European continent. Eventually, U5 descendants were themselves replaced by farmers and new immigrants to the continent; today they make up roughly 9 percent of European mitochondrial DNA. (In 2014, archeologists used improved carbon-dating techniques to show that Neanderthals may have gone extinct across the whole of Europe some 39,000 to 41,000 years ago.)

Gould believed we need to grasp and explore the consequences of a new perspective on the nature of history. As he wrote, "Neanderthal people were collateral cousins, perhaps already living in Europe while we emerged in Africa, and also contributing nothing to our immediate genetic heritage. In other words, we are an improbably and fragile entity, fortunately successful after precarious beginnings as a small population in Africa, not the predictable end result of a global tendency. We are a thing, an item of history, not an embodiment of general principles." There is no predestination in our survival and Neanderthal extinction.

Should we de-extinct Neanderthals in the future? It's obvious to any thinking person that a resurrected Neanderthal would be nothing like one from 30,000 years ago. We might give life to a genome, but the environment of a human womb—let alone an entirely new landscape and culture—would produce something very different from an original Neanderthal. No one can deny, however, that we would learn a lot, as we would from any de-extinction. Indeed, what makes Neanderthal de-extinction tantalizing is the opportunity to gain otherwise unattainable insight into one of the most fundamental questions in history, as Svante Pääbo has written: "Why did one type of human come to spread across the globe, replace all the other human forms and multiply to an extent that it influences much of the biosphere?"

＊　＊　＊

In the summer of 2014, the journal *Science* published the results of a genetic analysis of 169 ancient DNA samples collected from the Dorset people, an Arctic race that disappeared from the world 700 years ago. The Dorsets were Paleo-Eskimos, a shamanistic people whose livelihoods depended on hunting walrus and seal in eastern Canada and Greenland. No one really understood what happened to them, whether the Dorsets had assimilated into other populations of Eskimos or died out completely. Now genetics offered an important clue: the Dorset DNA samples showed a people who lived in isolation for some 4,000 years before going extinct in just a few decades, possibly weakened by inbreeding and climatic shifts that impacted their resources.

We might think of the extinction of a race like the Neanderthals as a relic of a far-gone age in our geological past, but they continued to take place throughout the twentieth century and even today. Consider that in 2010 the last member of an ancient tribe called the Bo died in India. This tribe had inhabited the Andaman Islands off India's eastern coast for 65,000 years and was believed to be one of the last peoples to have had contact with prehistoric people and remain unassimilated. In true extinctions like the Dorset and the Bo, entire ways of speaking about and conceptualizing the world disappear. The era of the Anthropocene is most often defined by human impact on the physical world, but as modernity and globalization sweep and transform communities around the planet, the nonbiological world of thought and language is impacted too. Modernization has become a kind of vortex of cultural assimilation that extinguishes *ways of being*, both animal and human. Interestingly, the modern mind-set, to which myself and the majority of people reading this book probably belong, presumes its superiority and centrality to history to such an extent that we don't really comprehend that there are different ways of existing in and thinking about the world.

The history of the idea of wilderness is a fascinating lens on this fact. How did prehistoric peoples think about nature and relate to species? It turns out that this is extraordinarily difficult for us to contemplate. According to the American philosopher Max Oelschlaeger, modernism—the period of history stretching all the way from the Renaissance to today and defined by science, capitalism, and a Judeo-Christian perspective on nature and time—obstructs our ability to inquire into the prehistoric experience of wilderness. Most likely, *they* didn't think about it at all. Prehistoric humans lacked a concept of "wilderness," of natural phenomena or spaces outside of the human domain. They almost certainly had a nondualistic relationship to the world in which there was no separation between mental and physical properties, and as a result, no division between human and natural phenomenon. These things were experienced seamlessly.

The Paleolithic mind, Oelschlaeger tells us, was an unmediated, prelinguistic, precultural mode of awareness. In his seminal work *The Idea of Wilderness*, Oelschlaeger wrote that prehistoric man likely "lacked reflective awareness of culture—that is, any conceptually clear realization that culture was a humanly initiated and sustained rather than an instinctive or natural mode of behavior—they thought of themselves as one with plants and animals, rivers and forests, as part of a larger, encompassing whole (which we would term a natural process or wild nature)." Once the agricultural revolution was underway, wilderness became an adversary to human survival, something that needed to be fought and controlled in order to serve human survival. Wilderness became a place outside of us, synonymous with wastelands, badlands, or hinterlands.

> Civilized people perpetuate the presupposition that prehistoric humans longed for paradise, some luxurious garden of easy living that would free them from travail and hunger. Because it assumes the categories and values of the modern world—indeed, the psychological profile of the modern mind—as absolutes, this argument invites deconstruction. From a modern perspective, a binary opposition, which can be neither critically nor empirically sustained, appears between archaic and modern culture and underlies the claim that so-called primitive people wanted to gain control over land and animals, and nature more generally. Most if not all evidence contradicts such a reading, and indicates that Paleolithic people lacked a concept of either a wilderness to escape or a civilization to seek. Only by holding our own categories in abeyance can we possibly understand the Paleolithic mind. The assumption that Paleolithic people were mere children in comparison to us, a later, adult phase of humanity, is dubious. So is the belief that the modern mind is the culmination of human intellectuality.

Our fervent belief in the objective superiority of the modern mode of existence perhaps biases us, for example, to assume that the reason Neanderthals didn't evolve the same technology as us is because they lacked intelligence, rather than they found a sustainable manner of living over 300,000 years. Modernists find it hard to imagine any desirable existence or definition of human except their own, argues Oelschlaeger. We think of prehistoric and indigenous peoples as living lives of extreme hardship and in a constant state of need, and that the modern mind arose out of an adaptive superiority rather than, say, an appetite for domination.

One great challenger to this sense of modern superiority was the French anthropologist Claude Lévi-Strauss. At a time when indigenous peoples were still commonly thought of as primitive, his book *The Savage Mind* presented an abundance of evidence that they in fact possessed a rich, complex, and sophisticated body of scientific knowledge about nature. To Lévi-Strauss, it was clear that various indigenous cultures had developed an extraordinary system of classification and systematics over thousands of years through methodical observation and tested hypotheses. For them, Lévi-Strauss wrote, plants and animals are not known as a result of their usefulness but are useful or interesting because they are known.

Pygmies in the Philippines, wrote Lévi-Strauss, could easily name 450 plants, seventy-five birds, twenty species of ants, and nearly every snake, fish, and insect. The Pinatubo of the Philippines had at least 600 named plants. "Several thousand Coahuila Indians never exhausted the natural resources of a desert region in South California, in which today only a handful of white families manage to subsist," wrote Lévi-Strauss. "They lived in a land of plenty, for in this apparently completely barren territory, they were familiar with no less than sixty kinds of edible plants and twenty-eight others of narcotic, stimulant or medicinal properties." The Hopi Indians named more than 350 plants and the Navajo at least 500. In Gabon, a French ethnobotanist compiled a list of 8,000 botanical terms from a half-dozen neighboring tribes. This knowledge wasn't necessarily limited to specific individuals who specialized in it. Children learned such things growing up, and the knowledge became the province of the tribe passed through the generations. It's this kind of intimate knowledge of nature, of relationships to the world, that disappear when a people or tribe go extinct, and equally so when a species goes extinct.

Lévi-Strauss published *The Savage Mind* in 1962, but decades later we resist the idea that for all of our scientific and technological achievements, modernity is not necessarily evidence of the forward progress of evolution and culture. Indeed, despite our quality of material existence and abundance of technology, the ecological problems of the twenty-first century have shown us that we don't understand nature much better—only very dif-ferently—than those that preceded us. Compared to even a hundred years ago, our lives are more distanced and divided from the natural world than ever before. Whereas the earth was once the origin of our history, lives, and source of survival, it has become an abstraction, a background to our every-day experience. Today, 54 percent of people live in urban environments, up from one-third in the 1960s, according to the World Health Organization. How many of us can claim an intimate knowledge of the animal and plant life around us? This might be one of the most telling reasons why stories about the disappearance of species often fail to capture our attention for more than a few moments. Their value is abstract to us. Even for those of us who purport to care very much, chances are the existence of species is not linked to our everyday experience or needs.

Our modern relationship to nature has its origins in a theistic worldview that arose in step with the agricultural revolution. God created the earth for humankind and our task as his servants is to create a New Jerusalem from the land, the Second Creation. In the twenty-first century, a great many of us have discarded this religiosity for a secular worldview that places its faith in scientism. Even science, however, can perpetuate the fractured relation-ship to nature by making us observers standing apart from the world even as we seek to understand how it works. Many modern environmentalists have sought to heal this breach between humans and nature by proposing new ecological worldviews that restore an ancient, lost connection to the earth. But their appeals, argues Oelschlaeger, however well intentioned, remain fundamentally Cartesian, a subset of the modern mind that believes nature is environment and therefore different from humans.

I began to wonder: If our ideas about nature are part of the multifari-ous roots of our current ecological crises, is it possible that philosophy can help us discover new ideas? As Holmes Rolston III has said, the question of what a species is is a scientific one to be answered by biologists, but the question of what duties we have toward it is an ethical one to be answered by philosophers.

✳   ✳   ✳

The more I learned about endangered species, the harder it became to comprehend the phenomenon of extinctions taking place today. Every case,
whether it was frogs, crows, or rhinos, was unique, complicated, and, relative to a specific context, without a quick fix. Some of the causes driving
these disappearances were cultural; others were political, economic, biological; and most were a mixture of all these things that together compounded
the challenges facing species and the question of how to save them. The term
"Sixth Extinction" felt even less satisfying to me than when I started out. It
was too vague, overwhelming. Each time I tried to define it and pin it down
to take a closer look, it seemed to slip away from me like one of those water
snake toys that you can never hold on to. And then one day I finally discovered the problem. I had been trying to grasp a hyperobject.

Hyperobjects are entities of such vast temporal and spatial dimensions
that they cannot fit into any previous definition of what an object is. Nonlo-
cal and massive, they span multiple human generations and even eons. You
can't see the end of them because they stretch so far into the future, you
can't really "see" them at all, and we can only witness aspects of them at
one time. Global warming is a hyperobject. So is nuclear radiation, and the
ecology of the Everglades. Evolution is a hyperobject. Also weather and the
oceans and whales. One feature of hyperobjects is the interconnectedness of
their different parts, and the more data we accumulate about all these parts,
the more complex the whole thing becomes. This may be why trying to understand North Atlantic right whales has consumed hundreds of researchers
for decades, and they are still discovering new questions to ask.

The concept of hyperobjects has been popping up in the mainstream
press in recent years, mostly in the context of climate change. In 2014, hyperobjects topped *Vogue*'s list of ten cultural things you need to know about,
which is somewhat surprising considering the concept's provenance in an
obscure philosophical movement. It was coined in 2010 by Timothy Morton,

a professor of English at Rice University, whose writing in philosophy, culture, and history is heady stuff. In 2013 Morton published a book called *Hyperobjects: Philosophy and Ecology After the End of the World*. The book is dense, and sometimes bewildering for anyone not schooled in academic philosophy. But it conveys a startling reality about our modern existence that, although strange, will feel familiar to many people. Morton says that we are living in a time of ecological crises that have revealed the reality of hyperobjects even as some of them, such as global warming and the Sixth Extinction, are challenging our political, intellectual, and ethical capacities.

A year after the book was published, I asked Morton whether he thought the enthusiastic reception for this new concept had something to do with people's relief at having a word to describe something they were already experiencing. "It helps people to cope if you can name it," said Morton. "It is a very intuitive idea. Hyperobjects are something that you can't feel or touch, but they are real. It's not just data. You can't see or touch extinction, but it's real." The difficulty is that hyberobjects are so big and complex we also can't really comprehend them with our individual brains. So it's not that hyperobjects showed up recently; some of them have been around forever. What's different now is that humans recently acquired the scientific tools to discern them.

Morton's work is heavily influenced by membership in a school of thought called "OOO," which stands for "object-oriented ontology." He didn't invent it but he is one of its most ardent cheerleaders. The OOO movement formally kicked off in 2010 at a symposium held at Georgia Tech in Atlanta. Among the participants was Graham Harman, a philosopher and professor at the American University in Cairo, who a decade earlier first used the term *object-oriented philosophy*. If you are an OOO believer, you must, according to Harman, hold true two assertions. First, that what he calls "entities of different scales" is the ultimate stuff of the universe. "Entities of different scales" is just another way of saying things or objects, anything and everything in existence: quarks, rocks, trees, polar bears, stars, pencils, Coke bottles, and computers. The second thing you must believe, writes Harman, is "that these entities are never exhausted by any of their relations or even by their sum of all possible relations." Understanding what Harman means by this second tenet of OOO requires knowing a little bit about the last 200 years of philosophy. As Harman explains it, ever since the German philosopher Immanuel Kant posited that humans can only experience things through our minds and senses—never directly—philosophers have been preoccupied with the idea that the world is only as real as it is accessible to humans.

What OOO posits is that objects are actually real whether or not humans can access them. Yes, they exist in relationship to humans, but they also exist in relationship to other objects. Harman put it this way in an interview: "The tree-in-itself is withheld from us not because we humans are specially tragic finite beings, but simply because we are objects at all. The wind has no more and no less direct access to the tree than I do." The way I began to think about OOO is that if the tree falls in the forest, it really falls whether someone saw it or not. The way Morton might put it, whether you believe in global warming (another sort of object) or not, sea levels really are rising.

What interested me is how philosophers are bringing the ideas in OOO to bear on the possibility of a new paradigm for critiquing our modern ideas about nature, and for hinting at a powerful nonanthropocentric perspective. Much of this has to do with throwing the idea of a split between humans and nature out the window. In interviews, Graham Harman cites the work of Bruno Latour, the French philosopher, who argues that the Anthropocene does not mean that nature is now subsumed by culture because humans have never been in a domain separate from nature. We are just one example of objects relating to other objects, nothing exists as more or less real than anything else.

This new interest in the nonanthropocentric realm has emerged in other disciplines too. Richard Grusin, a scholar of new media at the University of Wisconsin-Milwaukee and editor of the 2015 book *The Nonhuman Turn*, has written,

> Given that almost every problem of note that we face in the twenty-first century entails engagement with nonhumans—from climate change,

drought, and famine; to biotechnology, intellectual property, and privacy; to genocide, terrorism, and war—there seems no time like the present to turn our future attention, resources and energy toward the nonhuman broadly understood. Even the new paradigm of the Anthropocene, which names the human as the dominant influence on climate since industrialism, participates in the nonhuman turn in its recognition that humans must now be understood as climatological or geological forces on the planet that operate just as nonhumans would, independent of human will, belief, or desires.

For Morton, OOO gets rid of an idea of nature that defined both our human dominion over the natural world for millennia, and devalued nonhuman entities. As he puts it, "How do we transition from seeing what we call 'Nature' as an object 'over there'? And how do we avoid 'new and improved' versions that end up doing much the same thing . . . just in a 'cooler,' more sophisticated way? When you realize that everything is interconnected, you can't hold on to a single, solid, present-at-hand thing 'over there' called Nature."

The notion of separation between humans and nature is frequently embedded in the language of modern environmentalism. In his book *The End of Nature*, Bill McKibben writes, "An idea, a relationship, can go extinct, just like an animal or a plant. The idea in this case is 'nature,' the separate and wild province, the world apart from man to which he adapted, under whose rules he was born and died." In this sense, Morton happily cheers "the end of nature." He wants ecology without nature, a future in which we recognize there is no pristine wilderness, only history. In an ecology without nature, we're no longer the bouncer standing outside the club of existence, deciding who gets in or not, what has value or no value, what has rights or no rights at all. "Nature is a human construct with no bearing on reality whatsoever," Morton told me. "It's a thought-construct going back to the Middle Ages, and not only is it a toxic thought-construct that creates these false binaries between nature and culture, and modern and postmodern people . . . Nature *is* the problem, actual nature and how we got to that by demarcating social space from nonhumans and actually destroying earth."

I asked Morton if an ecology without nature provides a framework for thinking about human engineering in the Anthropocene, and whether it is natural or artificial. "Both ways of seeing and solving the problem are part of the problem," said Morton. "In an ecology without nature, instead of seeing the question as everything is natural so you can do whatever you like, or everything is artificial so you can never do anything you like, you are on the side of extreme hesitation. Let's not do it until we really think. Let's find ways to do biosynthetic stuff that is as kind to as many life-forms as possible. Let's be very tentative. And probably let's not geoengineer the hell out of earth." The goal of environmental ethics today shouldn't be about trying to make people care more about nature. It's a trap that environmentalists fall into to try to shout louder and make things sexier or more simplistic. "We stop trying to prove that the forest has intrinsic value," he told me. "Instead the question is: Do you love the forest? Are you completely entranced and hypnotized by it?"

OOO is a band of philosophers attempting to advance Western philosophy. Whether it should or will ever be appropriate for thinking about environmental policy making is questionable. But the litmus test for its worth, in my view, is its potential to foment a new way of thinking about species. OOO challenges us to consider them as objects that exist autonomously but experience us just as we do them. It requires us not only to acknowledge this reality but also to deeply consider how we interact with species. Some OOO thinkers, such as author and scholar Ian Bogost, for example, suggest we go further, to engage in the practice of "alien phenomenology," attempting to understand the experience and interiority of objects, no matter how incomprehensible or speculative an act this may be. What is it like to be a Florida panther swimming across the Caloosahatchee River? A North Atlantic right whale searching for food? The last northern white rhino? The wonder we might arouse in this process of contemplating such a thing could be the basis of respect and concern for species the world over.

CODA

Ultima Thule: Ends of the Earth

About 600 miles north of the northernmost coast of Norway in the Arctic Circle is an archipelago of frozen ice and earth called Svalbard. Since 2007 the government of Norway has sought to make Svalbard a UNESCO World Heritage Site, lauding it as one of the best-managed wilderness areas in the world and an example of authentic, unchanged nature. The largest and only populated island of the archipelago is Spitsbergen, and it's there, not far from the airport, that in 2008 the government blasted three chambers into the permafrost, creating the underground bunkers that make up the Svalbard Global Seed Vault. These spaces can hold millions of seeds representing the world's agricultural genetic diversity. According to Norway's Ministry of Agriculture and Food, an example of a recent shipment of samples sent to Svalbard included 575 types of barley, 5,964 kinds of wheat, and some heirloom red okra seeds from South Texas. The seed bank is one among hundreds of frozen repositories for biodiversity today, but its location is unique. For generations, the Norwegians have referred to Svalbard as *ultima thule,* "the ends of the earth." It's here that we are placing our collective faith, hoping that the reverberations of our environmental impacts won't be so great as to reach the ends of the earth. I've never been to Svalbard, but when I read about the Global Seed Vault, I realized that I already knew something about it, having read a rather obscure book called *A Woman in the Polar Night.*

This short book first appeared in German in 1938 and was translated into seven languages in subsequent years, including English in 1954. But it never gained much popularity outside Germany, even in the genre of Arctic travel and exploration. In 2010, however, the University of Alaska Press and Canadian publisher Greystone Books copublished the first English edition in fifty years. It sold a modest thousand copies, one of which I had bought after randomly discovering it on a bookshelf. The obscurity of *A Woman in the Polar Night* belies the book's singular revelations. In it, Christiane Ritter, an Austrian painter, wife, and mother, chronicles the year she spends living in Svalbard, where her contact with the remote Arctic wilderness transforms her as a person and leads her to discover, as she put it, the "consciousness of space."

Ritter was thirty-six years old when she went to Svalbard. Her husband had already given up his European life a few years earlier to live as a hunter and trapper in the archipelago, compelled by the solitude and beauty of the Arctic, which he first experienced in 1913 while working on a ship. He sent letters to Ritter in Austria that "told of journeys by water and over ice, of the animals and the fascination of the wilderness, of the strange illumination of one's own self in the remoteness of the polar night," the period of winter when the sun disappears for over four months and storms bombard the archipelago. Ritter's husband asked her to visit him. "It won't be too lonely for you because at the northeast corner of the coast, about sixty miles from here, there is another hunter living, an old Swede. We can visit him in the spring when it's light again and the sea and fiords are frozen over." Intrigued by his letters, Ritter decided to join him with the intention of reading "thick books in the remote quiet and, not least, sleep to my heart's content."

In the early twentieth century, society considered the wilderness inhospitable to women. This was particularly true of the Arctic, where the extreme climate was thought to be too physically and emotionally punishing for females. It's generally believed that no woman set foot on Svalbard until 1838, and none lived there permanently or stayed through the winter until Ritter made up her mind to do it.

She arrived by ship in July of 1934, a time of year when the sun never sets, to a landscape drenched in overcast light, water, mist, and rain. Once upon a time, hundreds of millions of years ago, Svalbard was located south of the equator until continental drifts pushed it northward. Around 60 million years ago it reached the latitude of today's southern Norway and was covered by swamplands. When the ice ages started a couple million years ago, Svalbard began to look like what it does today. Roughly the size of Ireland and one of the northernmost landmasses in the world, its western portions contain high mountains while the rest of the archipelago is made up of flat moorlands, tundra, canyons, fjords, glaciers, moraines, and ancient rock, some containing minerals billions of years old.

Ritter's home for the next year was a small hut with a soot-spewing stove for heat, on a promontory with banks that dropped into the sea. Stone and animal skeletons, an "arid picture of death and decay," surrounded the hut. In the beginning, Ritter couldn't seem to get over the bleakness of her surroundings, and the lack of structure to the days. "One day melts into the next, and you cannot say this is the end of today and now it is tomorrow and that was yesterday. It is always light, the sea is always murmuring, and the mist stands immovable as a wall around the hut. We eat when we are hungry; we sleep when we are tired." The lack of routines or clocks began to deprogram Ritter from her former life, initiating her into a mind-set that became exquisitely attuned to the ever-changing environment around her. Svalbard's wilderness was not pristine: between 1699 and 1778 some 8,500 whales were killed by Dutch whalers. In the nineteenth century, bowhead whales had been hunted almost to extinction, and only some fin, beluga, and killer whales remained; walrus populations had been nearly wiped out. But because of its remote, harsh conditions, it was a place where one could still go months if not years without any contact with other humans or signs of civilization.

Ritter described the day in October when the sun disappeared, not to be seen again for 132 days. The onset of night resulted in a decisive change in the human mood, wrote Ritter, "when the reality of the phenomenal world dissolves, when men slowly lose all sense of fixed points, of impulses from the external world." A few weeks later, a storm began to rage while Ritter's husband was away hunting and she was alone in the hut. The gale buried the front door in a snow barrier ten yards wide. The paraffin for the lamp ran out and Ritter was left in the dark. The noise from the storm was so loud it was like a deafening rumble. But in order to get to the coal to light the fire and prevent freezing to death, Ritter had to get fuel. She dug through the snow and crawled on the ground to prevent being blown away into the pitch-blackness, gathered her coal, and made it back to the hut. She performed the feat again and again as the storm raged for nine days and nights. "After a while my hands begin to tremble," she wrote. "I catch myself creeping softly about the hut, doing all my jobs slowly with measured movements, as though trying not to attract the attention of the raging deity outside."

When the storm broke, Ritter was a different person, humbled and in awe of the world. She took her skis into the quiet snow:

The power of this worldwide peace takes hold of me, although my senses are unable to grasp it. And though I were unsubstantial, no longer there, the infinite space penetrates through me and swells out, the surging of the sea passes through my being, and what was once a personal will dissolves like a small cloud against the inflexible cliffs. I am conscious of the immense solitude around me. There is nothing that is like me, no creature in whose aspect I might retain a consciousness of my own self; I feel that the limits of my being are being lost in this all-too-powerful nature, and for the first time I have a sense of the divine gift of companionship.

In the book, Ritter predicts that in centuries to come men will go to the Arctic "as in biblical times they withdrew to the desert, to find the truth again." This truth is the fact of humanity's origins in wilderness, and our wild longings "stronger than all reason and all memory" to return to places where the human scale is annihilated by the powerful mystery of existence. The seeping of our consciousness into infinite space is our ultimate salvation, says Ritter, putting our human reasoning into perspective. Her book is full of images evoking the infinitesimal size of humans against the landscape, the image of a person like a "tiny piece of coal" against phenomena so powerful we have to stretch our mind and spirit to try to comprehend them. Svalbard is a place where she tells us we can witness the "unfathomable gulf between human magnitudes and eternal truth."

Ritter might never have been able to predict the impact that humans would have on the Arctic in the form of climate change in the decades to come, and what she would think about the doomsday vault in Svalbard we can only guess; she passed away in 2000 at the age of 103. Until recently, Svalbard appeared to have escaped some of the larger climatic changes that have affected other parts of the Arctic so dramatically. Unlike neighboring Greenland, where climate change in recent years has caused as much as 97 percent of the surface of the ice sheet to melt during the summers, regional climate models in Svalbard showed no acceleration of surface melt on the island from the 1970s to 2012, according to a report in the *Journal of the European Geosciences Union*. But this may actually have been a fluke *result* of climate changes. Atmospheric circulation changed during that period, bringing northerly flows over the island during the summers and helping to keep the temperatures moderate. Then in 2013, these conditions switched: southwesterly flows over the territory caused record-level melts. This hasn't been the only sign that changes are amuck in Svalbard. In the Barents Sea to the west, giant phytoplankton blooms are beginning to appear in the fall as pack ice takes longer to form and sunlight warms the waters. Researchers from the Norwegian Polar Institute are noticing changes in polar bear behavior. In the 1980s they began counting female polar bears traveling to give birth in Kongsøya, in the eastern part of the archipelago. In the mid-1980s they counted fifty; in 2009 there were twenty-five, and in 2012 the number was down to five. "We see that there is a clear link between sea ice extent and females' opportunity to reach their hibernating areas," researcher Jon Aas told Norway's largest newspaper, *Aftenposten*.

I hope to visit Svalbard some day to experience what Ritter called the "consciousness of space," and perhaps glean some of the truth Ritter discovered there. No one could deny that the truth is more complicated than when she first stepped ashore. It involves doomsday vaults and vanishing animals. It's a story about one species that has multiplied so quickly and become so artful that it has the power to influence, for better or worse, the evolutionary processes that produced it, and to ultimately determine the fate of other species. Science and technology can make us feel that the gulf between human magnitudes and eternal truth has been bridged. We can peer into genomes, locate the beginning of the Big Bang on the outer edges of space, travel down the Grand Canyon via Google Earth, and clone animals. I was able to take a virtual tour of Ritter's hut online. But even as we assume a mythic God-like perspective on the universe, the majority of us who live in the modernized world have no idea how to survive away from civilization. How quickly would we revert to primitive awe if directly confronted with nature's power, alone in the Arctic night? What should we do to save that place and its creatures, and its ability to transform us?

In recent years, experts have been telling us that the epoch of the Anthropocene is upon us. Our influence over nature as a result of climate change is so pervasive that wilderness itself has gone extinct. Two-thirds of the planet's land area is devoted to supporting human activities. Even Aldo Leopold, a founding father of wildlife ecology, anticipated this way back in 1933, when he wrote that it was too late to curtail the influence of human occupancy on nature. "Every head of wild life still alive in this country is already artificialized, in that its existence is conditioned by economic forces. . . . The hope of the future lies not in curbing the influence of human occupancy—it is already too late for that—but in creating a better understanding of the extent of that influence and a new ethic for its governance."

Should the future of conservation involve more management of nature, or take the form of explicitly engineering or de-extincting species? I think this question merits careful consideration before humanity relinquishes the existence of wild places and wild things in the world. In that future of human-dominated landscapes and geoengineering, humility may be what we really stand to lose, the ability to be reminded of our mortal magnitude. As wilderness and living things untouched and independent of human influence become rarer, so does the overwhelming evidence that we are not the culmination of an evolutionary tendency but rather, as Stephen Jay Gould said, its lucky byproduct.

There may be other important reasons to be extraordinarily cautious rather than enthusiastic about humans trying to geoengineer a brighter future. The more we attempt to control nature, the more we find out that the earth's biological systems are interconnected with astonishing complexity; we can't predict the myriad ways our attempts will have unintended consequences or even unleash some of its more extreme powers: natural disasters, disease, the loss of ecosystems, and further extinctions. Ultimately, I have heard few cases for de-extinction that present a clear and compelling ethical argument that justify the risks; many of them appear to me to reflect aspects of the current preoccupations of modernity with technology, mortality, and the end times. At moments arguments for de-extinction feel like a kind of spiritual denial that evolution and extinction are two sides of the same coin, and so long as we choose to fuel our survival as a species through the exploitation of natural resources, extinctions may well be the cost of our appetite for progress. Until we make space for other species on earth, it won't matter how many animals we resurrect, there won't be many places left for them to exist.

ACKNOWLEDGEMENTS

I would like to start by thanking my friend Tom de Zengotita, which will surprise the hell out of him. But if it were not for attending his own book launch I might never have ended up in graduate school, and if not for his suggestion that I write a book and his introduction to his agent, I would not have gotten this far. My heartfelt thanks to an amazing agent, Michelle Tessler, who took a very obscure idea and helped me polish it and give it depth through her encouragement and enthusiasm. My greatest appreciation to Elisabeth Dyssegaard for seeing the potential in these stories early on, her supremely thoughtful editing, and her wonderful ideas and warmth. The completion of this book benefited from the generous support and confidence of the Alfred P. Sloan Foundation's Program for the Public Understanding of Science, Technology, & Economics, and in particular Doron Weber. Sincere gratitude to Alan Bradshaw for his careful stewardship of the manuscript, Bill Warhop and Carol McGillivray for their terrific copyediting and patience, Laura Apperson for her cheer and assistance, and David Baldeosingh Rotstein for the book's cover design.

Journalists would not amount to much without the expertise and generosity of their sources. This is particularly true in my case as I required much hand-holding while learning about the science and history of conservation. Seeing the commitment of so many brilliant individuals to the cause of science and conservation and ethics was humbling. Thank you for your time, thoughtfulness, and willingness to be harassed, in no particular order: Bill Newmark, Kim Howell, Jenny Pramuk, Andy Odum, Charles Msuya,

Ché Weldon, Kurt Buhlmann, Eric Katz, Bryan Norton, Holmes Rolston III, Brad White, Phillip Hedrick, Dave Onorato, Darrell Land, Nate Greve, Laurie Macdonald, Steve Williams, Chris Belden, Rocky McBride, Michael Kinnison, Craig Stockell, Michael Collyer, Scott Caroll, John Pittinger, Kevin Rice, Małgorzata Ożgo, Michael Barkham, Andrew Pershing, Bob Kenney, Brad White, Clay George, Katie Jackson, Brenna McLeod, Tom Pitchford, Oliver Ryder, Tracey Heatherington, Janet Chernela, Joanna Radin, Julie Feinstein, Thom van Dooren, Susan Ellis, Jeanne Loring, Julien Delord, Mohammed Doyo, George Church, Hank Greely, Ben Novak, Garrie Landry, Joel Greenberg, Metin Erin, John Gerstle, Peter Hawkes, Chris Wemmer, Brian Gratwicke, Tim Herman, Jim Hain, Ed Lyle and Tim Morton. I must single out Roy McBride, who was particularly gracious with his time and insights despite his ambivalence toward my endeavor. Also, my many thanks to George Amato for his wonderful conversations and thoughts. My heartfelt gratitude to Kes and Frasier Smith in Nairobi for their remarkable generosity and openness and for their home staff, in particular Lucy for teaching Joaquín to crawl.

I owe a great deal to Sandro Stille, whose graduate course "Covering Ideas" was not only where I first wrote about the Kihansi frogs but confirmed for me the belief that ideas are the legitimate purview of journalism; thank you for offering support and friendship as I navigated a career. Thank you also to Bruce Porter and Billy Gorta, who taught me everything I know about reporting, particularly fires, riots, and homicides. *Asante sana* to Marla Jaksch, who is an inspiration, Warcheera Said, Musa Binzayed Hussein, Ken and Monica Okoth, and Micah Filipho for taking me under their wings in Dar es Salaam. Thank you to Ross Robertson, for his humor, his friendship over many years, and his hospitality in Santa Cruz. My affection to heroic moms Laura Snelgrove, Mary Kate Wise, Janelle Wilson, Alice Tang, Sara Lupita Olivares, and Jenny Bohrman. Also my cement-pounding journo friends: Danny Gold, Neil Munshi, Mary Cuddehe, Matt Lysiak, Bob McDonald, Nicholas Phillips, Gianna Palmer, Alex Halperin, Denver Nicks, Warzer Jaff and Bill Farrington. To Kaitlin Bell Barnett, Lygia Navarro, Deborah Jian Lee, Morgen Peck and Susana Ferreira for their fantastic conversation and moral support. Many thanks to Chanelle Elaine and Jillian Cambell. Thanks to Tom Peter and Emma Piper-Burket for road-tripping and babysitting in White Sands, and the whole Anderson Family for their friendship and generosity in Santa Fe.

Life wouldn't be as exciting without the feisty and passionate Malcolm Wyer—thank you for your advice and unconditional friendship. Thank you to Bob and Janet Miller, for providing a home away from home to write in Anna Maria not once but twice, and for your amazing love and encouragement over many years. I owe much to my dad, Rory O'Connor, for his ceaseless, unabashed interest in the world that has been the spirit behind my work. My mom, Katherine Miller: thank you for your radiant humanity and giving me your love of the outdoors. Robert Heinzman, for taking care of my mom and for your encouragement and support of this book at critical moments. My gratitude to my Seattle family, George, Margaret, Maureen, and Suraya Parker, for your warm embrace. And much love and thanks to my beautiful sister, Jane O'Connor in Dublin: together we'll carry on the family name for glory and treasure.

Last, my deepest love and thanks to Bryan Parker, whose big-heart and characteristic grace made this endeavor possible. I hope our adventures are endless.

NOTES

INTRODUCTION

"extinction, it has been said" Kent H. Redford et al., "What Does It Mean to Successfully Conserve a (Vertebrate) Species?," *BioScience* 61, no. 1 (January 2011): 39–48, doi:10.1525/bio.2011.61.1.9.

"A society that is habituated" Ronald R. Swaisgood and James K. Sheppard, "The Culture of Conservation Biologists: Show Me the Hope!," *BioScience* 60, no. 8 (September 2010): 626–30, doi:10.1525/bio.2010.60.8.8.

"extinction debt" Fangliang He and Stephen P. Hubbell, "Species-Area Relationships Always Overestimate Extinction Rates from Habitat Loss," *Nature* 473, no. 7347 (May 19, 2011): 368–71, doi:10.1038/nature09985.

"A meeting of conservation biologists" M. J. Costello, R. M. May, and N. E. Stork. "Can We Name Earth's Species Before They Go Extinct?" *Science* 339, no. 6118 (January 25, 2013): 413. doi:10.1126/science.1230318.

"10 percent of the earth's land" Buckley, Robert M., Patricia Clarke Annez, and Michael Spence, eds. *Urbanization and Growth*. The World Bank, 2008. http://elibrary.worldbank.org/doi/book/10.1596/978-0-8213-7573-0.

"Caribou have lost" Schaefer, James A. "Long-Term Range Recession and the Persistence of Caribou in the Taiga." *Conservation Biology* 17, no. 5 (2003): 1435.

"not replaceable without depreciation" Robert Elliot, "Faking Nature," *Inquiry* 25, no. 1 (January 1, 1982): 81–93, doi:10.1080/00201748208601955.

CHAPTER 1: AN ARK OF TOADS

"the area to be lost" Ekono Energy. *Kihansi Hydroelectric Project Environmental Assessment*. Environmental Assessment. Kihansi Hydroeletric Project. Nordic Development Fund, July 31, 1991.

"reliable power is so important" Paul Romer, "For Richer, for Poorer," *Prospect Magazine*, February 2010. http://www.prospectmagazine.co.uk/features/for-richer-for-poorer.

"esthetic, ecological, educational, recreational, and scientific value" "Endangered Species I Laws & Policies I Endangered Species Act," Signed into law, December 28, 1973. U.S. Fish and Wildlife Service, http://www.fws.gov/endangered/laws-policies/.

"made to serve" Bryan G. Norton, *Why Preserve Natural Variety?*. Princeton, N.J.: Princeton University Press, 1990, 195.

"last man" Richard Sylvan (Routley). "Is There a Need for a New, an Environmental, Ethic?" In *Environmental Ethics: An Anthology,* edited by Andrew Light and Holmes Rolston III, 1 edition., 49. Malden, MA: Wiley-Blackwell, 2002.

"Human interests and preferences" Ibid., 52.

"intrinsic value" For discussion and background on the idea of intrinsic value in environmental ethics, see J. Baird Callicott. "Rolston on Intrinsic Value: A Deconstruction," Environmental Ethics, Vol. 14, No. 2 (1992): 129–43, doi:10.5840/enviroethics199214229.

"the world, we are told" John Muir and Peter Jenkins. *A Thousand-Mile Walk to the Gulf.* Boston: Mariner Books, 1998, 136.

"theoretical quest" J. Baird Callicott. "Rolston on Intrinsic Value: A Deconstruction." *Environmental Ethics* 14, no. 2 (1992): 129–43. doi:10.5840/enviroethics199214229., 129.

"Perhaps there can be no science" Holmes Rolston III. "Value in Nature and the Nature of Value." In Philosophy and the Natural Environment, edited by Robin Attfield and Andrew Belsey :13–30. Royal Institute of Philosophy Supplement. University of Wales, Cardiff: Cambridge University Press, 1994, 29.

"an adventure in what it means" Holmes Rolston III. "PL: 345 Environmental Ethics," Fall 2002. http://lamar.colostate.edu/~rolston/345-SYL.htm.

"Our efforts at conservation" Stephen Jay Gould, *An Urchin in the Storm: Essays About Books and Ideas* (New York: W. W. Norton, 1988), 21.

"we shall finally understand" Ibid., 21.

"The species defends a particular form of life," "Value in Nature and the Nature of Value." In Philosophy and the Natural Environment, edited by Robin Attfield and Andrew Belsey: 13–30. Royal Institute of Philosophy Supplement. University of Wales, Cardiff: Cambridge University Press, 1994, 21.

"form of life is unique, warranting respect" "World Charter for Nature, 48th Plenary Meeting," United Nations A/RES/37/7, October 28, 1982, http://www.un.org/documents/ga/res/37/a37r007.htm.

"intrinsic value of biological diversity" "Preamble," Convention on Biological Diversity, June 5, 1992. http://www.cbd.int/convention/articles/default.shtml?a=cbd-00.

"he named academic philosophy" J. Baird Callicott, *Beyond the Land Ethic: More Essays in Environmental Philosophy* (Albany: State University of New York Press, 1999), 42.

"continue to dance with Cartesian ghosts" Bryan G. Norton, "Epistemology and Environmental Values," *Monist,* April, 1992, 224.

"From an environmental professional's perspective" John Lemons. "Nature Diminished or Nature Managed: Applying Rolston's Environmental Ethics in National Parks." In *Nature, Value, Duty: Life on Earth with Holmes Rolston, III,* edited by Christopher J. Preston and Wayne Ouderkirk, Springer Netherlands, 2010, 212.

"There is no evidence for" Michael Soulé, "The 'New Conservation,'" *Conservation Biology* 27, no. 5 (October 2013): 895–97, doi:10.111/cobi.12147.

"Is it worth" Author's field notes, Tanzania, 2010.

CHAPTER 2: TRACKING CHIMERAS IN THE FAKAHATCHEE STRAND

"The exact range of the form" Boston Society of Natural History, *Proceedings of the Boston Society of Natural History,* vol. 28 (Ulan Press, 2011), 235.

"During his younger days" Donald G. Schueler, *Incident at Eagle Ranch: Predators as Prey in the American West* (Tucson: University of Arizona Press, 1991), 177.

"The wolf seldom used the same trail twice" Roy T. McBride, The Mexican Wolf (Canis Lupus Baileyi): A Historical Review and Observations on Its Status and Distribution: A Progress Report to the U.S. Fish and Wildlife Service. U.S. Fish and Wildlife Service, 1980, 33.

"Almost a year had passed" Ibid., 33.

"I set a trap" Ibid., 33.

"Not many" Roy T. McBride, "Three Decades of Searching South Florida for Panthers," (presentation at the Proceedings of the Florida Panther Conference, Fort Myers, Florida, November 1, 1994). http://www.panthersociety.org/decades.html.

"I was amazed to find them" Ibid.

"I don't like those damn things" Craig Pittman, "Young Florida Panther Shot Dead on Big Cypress Preserve," *Tampa Bay Times*, December 9, 2013, http://www.tampabay.com/news/environment/wildlife/panther-shot-dead-on-big-cypress-preserve/2156228.

"The average sperm count" U.S. Fish and Wildlife Service. *Final Environmental Assessment: Genetic Restoration of the Florida Panther.* Gainesville, Florida, December 20, 1994, 3.

"classic example of what happens" David Maehr, *The Florida Panther: Life and Death of a Vanishing Carnivore* (Washington, DC: Island Press, 1997), xi.

"quick fix to a complex problem" Ibid., xi.

"reinstate gene flow" U.S. Fish and Wildlife Service. *Final Environmental Assessment: Genetic Restoration of the Florida Panther.* Gainesville, Florida, December 20, 1994, 5.

"human-caused isolation" Ibid., 5.

"genetic augmentation" David Maehr, *The Florida Panther: Life and Death of a Vanishing Carnivore* (Washington, DC: Island Press, 1997), 204.

"genetic restoration" Ibid., 204.

"anthropogenic hybridization" Fred W. Allendorf, Paul A. Hohenlohe, and Gordon Luikart. "Genomics and the Future of Conservation Genetics." *Nature Reviews Genetics* 11, no. 10 (October 2010): 697–709. doi:10.1038/nrg2844.

"The possibility that a subspecies carries" Stephen J. O'Brien and Ernst Mayr, "Bureaucratic Mischief: Recognizing Endangered Species and Subspecies," *Science* 251, no. 4998 (March 8, 1991): 1187 (2).

"thing of immortal make, not human" Matt Kaplan, *The Science of Monsters: The Origins of the Creatures We Love to Fear* (New York: Scribner, 2013), 34.

"I don't think ranchers should" Murray T. Walton, "Rancher Use of Livestock Protection Collars in Texas," In Proceedings of the Fourteenth Vertebrate Pest Conference 1990, 80, 1990., 277.

"I've done it all" Rick Bass, *The Ninemile Wolves* (1992; repr., Boston: Mariner Books, 2003), 79.

CHAPTER 3: EXUBERANT EVOLUTION IN A DESERT FISH

"the domestic races of many animals and plants" Charles Darwin. *The Origin of Species by Means of Natural Selection, or the Preservation of Favoured Races in the Struggle for Life*, 6th Edition. (New York: Cambridge University Press, 2009),12.

"The most simple of these is the biological concept" Ernst Mayr. "What Is a Species, and What Is Not?" *Philosophy of Science* 63, no. 2 (June 1996): 262.

"the phylogenetic concept" Paul-Michael Agapow et al., "The Impact of Species Concept on Biodiversity Studies," *Quarterly Review of Biology* 79, no. 2 (June 2004): 161, doi:10.1086/383542.

"In 2004, researchers published" Ibid., 161.

"lineages of ancestral descent" E. O. Wiley, "The Evolutionary Species Concept Reconsidered," *Systematic Biology* 27, no. 1 (March 1, 1978): 17–26, doi:10.2307/2412809.

"Assiduous collecting up cliff faces" Niles Eldredge, *Reinventing Darwin: Great Evolutionary Debate* (London: Weidenfeld & Nicolson, 1995), 95.

"most obvious and gravest objection" Charles Darwin. *On the Origin of Species by Means of Natural Selection, or the Preservation of Favoured Races in the Struggle for Life.* (London: W. Clowes and Sons, 1859), 280.

"rejecting that which is bad" Ibid., 84.

"we see nothing of these slow changes" Ibid., 84.

"thirty to sixty generations" David A. Reznick, Heather Bryga, and John A. Endler. "Experimentally Induced Life-History Evolution in a Natural Population." *Nature* 346, no. 6282 (1990): 357.

"a year of fire" Oscar Edward Meinzer and Raleigh Frederick Hare. *Geology and Water Resources of Tularosa Basin, New Mexico.* 343. (Washington, DC: United States Geological Survey, Department of the Interior, 1915), 23.

"valley was filled with flames" Ibid., 23.

"Because Lost River and Mound Spring populations" Michael L. Collyer et al., "Morphological Divergence of Native and Recently Established Populations of White Sands Pupfish (Cyprinodon tularosa)," *Copeia* 2005, no. 1 (2005), 9.

"might be too costly a gamble" Michael L. Collyer, Jeffrey S. Heilveil, and Craig A. Stockwell, "Contemporary Evolutionary Divergence for a Protected Species Following Assisted Colonization," *PLoS ONE* no. 6(8): e22310. doi:10.1371/journal.pone.0022310. (August 2011), 6.

"might best be viewed as evolutionary experiments" Ibid., 5.

"preserve species as dynamic entities" Richard Frankham, Jonathan D. Ballou, and David A. Briscoe, *Introduction to Conservation Genetics,* 2nd ed. (Cambridge, UK; New York: Cambridge University Press, 2010), 119.

"the greatest contribution that evolutionary rate" A. P. Hendry and M. T. Kinnison. "The Pace of Modern Life: Measuring Rates of Contemporary Evolution," *Evolution: International Journal of Organic Evolution* 53, no. 6 (1999): 1650.

"We challenge conservation biologists" Craig A. Stockwell, Andrew P. Hendry, and Michael T. Kinnison, "Contemporary Evolution Meets Conservation Biology," *Trends in Ecology & Evolution* 18, no. 2 (2003): 99.

CHAPTER 4: MYSTERIES OF THE WHALE CALLED 1334

"most whale calves will likely" Philip K. Hamilton, Amy R. Knowlton, and Marilyn K. Marx, "Right Whales Tell Their Own Stories: The Photo-Identification Catalog," In *The Urban Whale: North Atlantic Right Whales at the Crossroads,* ed. Scott D. Kraus and Rosalind M. Rolland, (Cambridge, MA: Harvard University Press, 2007), 96.

"And every day we saw whales" Frederick W. True. "The Whalebone Whales of the Western North Atlantic, Compared with Those Occuring in European Water; With Some Observations On the Species of the North Pacific." In *Smithsonian Contributions to Knowledge,* Vol. 33. (Washington, DC: The Smithsonian Institution, 1904), 22.

"Albatrosses appear to challenge" E. Milot, H. Weimerskirch, P. Duchesne and L. Bernatchez. "Surviving with Low Genetic Diversity: The Case of Albatrosses." *Proceedings of the Royal Society B: Biological Sciences* 274, no. 1611 (March 22, 2007): 785. doi:10.1098/rspb.2006.0221.

"morning mowers, who side by side" Herman Melville, *Moby Dick: Or the Whale* (London: Modern Library, 1992), 396.

"The climate-driven changes in ocean" Charles H. Greene et al., "Impact of Climate Variability on the Recovery of Endangered North Atlantic Right Whales, "*Oceanography* 16, no. 4 (2003): 100. doi.org/10.5670/oceanog.2003.16

"Ultimately, our ability to assess" Ibid., 102.

CHAPTER 5: FREEZING CROWS

"The ability to examine" Fred W. Allendorf, Paul A. Hohenlohe, and Gordon Luikart. "Genomics and the Future of Conservation Genetics." *Nature Reviews Genetics* 11, no. 10 (October 2010), 697. doi:10.1038/nrg2844.

"acquired evolutionary responsibility" L. T. Evans, "Sir Otto Frankel: Biographical Memoirs," Australian Academy of Science, 1999.

"Neither our pre-agricultural ancestor" Otto H. Frankel, "Genetic Conservation: Our Evolutionary Responsibility," *Genetics* 78, no. 1 (1974): 54.

"There is a benefit in maintaining genetic diversity" Janet Chernela, "A Species Apart: Ideology, Science, and the End of Life," In *The Anthropology of Extinction: Essays on Culture and Species Death*, ed. Genese Marie Sodikoff (Bloomington and Indianapolis: Indiana University Press, 2012), 30.

"Cryopreservation of gametes and embryos" Professor George Amato, Professor Howard C. Rosenbaum, and Professor Rob DeSalle. *Conservation Genetics in the Age of Genomics.* (New York: Columbia University Press, 2009), 61.

"It took all one's scientific ardor" Lyle Rexer et al., Carl E. Akeley. *In Brightest Africa* (Garden City: Doubleday, 1923), 229.

"the life that the organismic individual has" Holmes Rolston III, *Genes, Genesis, and God: Values and Their Origins in Natural and Human History* (Cambridge, UK: Cambridge University Press, 1999), 42.

"Cells can be kept frozen" Andrea Johnson, "Preserving Hawaiian Bird Cell Lines," *Animals & Plants* (blog), San Diego Zoo, November 7, 2008, http://blogs.sandiegozoo.org/2008/11/07/preserving-hawaiian-bird-cell-lines/.

"One of the rewarding things" Ibid.

"to bawl, bleat, squeal, cry" US Fish and Wildlife Service, "Revised Recovery Plan for the 'Alala (*Corvus hawaiiensis*)," Portland, Oregon, January 27, 2009. http://www.fws.gov/pacific/eco services/documents/Alala_Revised_Recovery_Plan.pdf

"It would be difficult to imagine" Mark Jerome Walters. *Seeking the Sacred Raven: Politics and Extinction on a Hawaiian Island.* 2nd ed. (Washington D.C.: Island Press, 2006) 52.

"The arrogance implied" Mark Jerome Walters, *Seeking the Sacred Raven: Politics and Extinction on a Hawaiian Island,* 2nd ed. (Washington, DC: Island Press, 2006), 145.

"They were kind of like the kings and queens of the forest" Thom van Dooren. "Authentic Crows: Identity, Captivity and Emergent Forms of Life." *Theory, Culture and Society,* forthcoming.

"Conserving species is" Ibid.

"delicately interwoven ways of life" Thom van Dooren. "Banking the Forest: Loss, Hope and Care in Hawaiian Conservation." In *Defrost: New Perspectives on Temperature, Time, and Survival,* edited by Joanna Radin and Emma Kowal, forthcoming.

"All cryo-technologies" Ibid.

"If the death of a single crow signals" Thom van Dooren, *Flight Ways: Life and Loss at the Edge of Extinction* (New York: Columbia University Press, 2014), 142.

a quote the Frozen Ark "The Future of the Frozen Ark." *The Frozen Ark: Saving the DNA of Endangered Species.* http://www.frozenark.org/future-frozen-ark. accessed December 6, 2014.

"With its biblical reference" Tracey Heatherington, "From Ecocide to Genetic Rescue: Can Technoscience Save the Wild?," In *The Anthropology of Extinction: Essays on Culture and Species Death*, ed. Genese Marie Sodikoff (Bloomington and Indianapolis: Indiana University Press, 2012), 40.

"Loss of genetic diversity" Bryan G. Norton, *Why Preserve Natural Variety?* (Princeton, NJ: Princeton University Press, 1990), 260.

"keep the endangered forest" George Amato. "Moving Toward a More Integrated Approach." In *Conservation Genetics in the Age of Genomics,* edited by George Amato, Howard C. Rosenbaum, Rob DeSalle, and Oliver A. Ryder. (New York: Columbia University Press, 2009), 36.

"I could not limit myself" Walters, *Seeking the Sacred Raven,* 110.

CHAPTER 6: METAPHYSICAL RHINOS

"two bulls loomed out" Ian Player. *The White Rhino Saga.* 1st edition. (New York: Stein and Day, 1973), 17.

"'real essentialism' and 'three-dimensional individualism'" Julien Delord. "Can We Really Re-Create an Extinct Species by Cloning?" In *The Ethics of Animal Re-Creation and Modification: Reviving, Rewilding, Restoring,* edited by Markku Oksanen and Helena Siipi. (New York, Palgrave Macmillan, 2014), 28.

"pocket-sized Venus" Alan Root. *Ivory, Apes & Peacocks: Animals, Adventure and Discovery in the Wild Places of Africa.* (London: Chatto & Windus, 2012), 259.

"Kes is a formidable woman" Douglas Adams and Mark Carwardine, "Last Chance to See," (repr., New York: Ballantine Books, 1992), 84.

"huge herds moving across" Kes Smith, ed., *Garamba: Conservation in Peace and War,* forthcoming.

"has run in parallel with" Alan Root, *Ivory, Apes & Peacocks: Animals, Adventure and Discovery in the Wild Places of Africa* (London: Chatto & Windus, 2012), 299.

CHAPTER 7: REGENESIS OF THE PASSENGER PIGEON

"feathered tribes" Clark Hunter, ed., *The Life and Letters of Alexander Wilson,* Vol. 154 (Philadelphia: Memoirs of the American Philosophical Society, 1983), 100.

"While others are immersed" Ibid., p. 106.

"great Author of the Universe" Ibid., p. 269.

"There are such prodigious numbers" A. W. Schorger, *The Passenger Pigeon: Its History and Extinction* (Caldwell, NJ: Blackburn Press, 2004), 11.

"unearthly" A. W. Schorger, "The Great Wisconsin Passenger Pigeon Nesting of 1871," *Passenger Pigeon: Monthly Bulletin of the Wisconsin Society of Ornithology* 1, no. 1 (February 1939): 31.

"When such myriads" Schorger, *Passenger Pigeon,* 54.

"arose a roar, compared with" Ibid., 189.

"The slaughter was terrible" Ibid.

"The passenger pigeon needs no protection" Ibid., 225.

"If the world will endure" Ibid., 208.

"In due course, the day will come" Mark V. Barrow Jr., *Nature's Ghosts: Confronting Extinction from the Age of Jefferson to the Age of Ecology* (Chicago: University of Chicago Press, 2009), 127.

"This species became extinct" Schorger, *Passenger Pigeon,* 230.

"culminating effort of Nature" William Beebe, *The Bird: Its Form and Function* (1906; repr., Ulan Press, 2012), 17.

"Let us beware of needlessly" Ibid., 18.

"Since my childhood days" George Landry, "The Final Tale of a Passenger Pigeon Named 'George,'" Exotic Dove website, accessed September 2013 www.exoticdove.com/P_pigeon /George_3.html.

"Like a dodo bird" Ben Novak, "Flights of Fancy: A Tiny Tube of Clear Liquid," Project Passenger Pigeon, n.d., http://passengerpigeon.org/flights.html.

"deep ecological enrichment" Ryan Phelan, "About TEDxDeExtinction and TED," Revive & Restore, Long Now Foundation, accessed December 6, 2014, http://longnow.org/revive/events /tedxdeextinction/about/.

"genetic rescue" "What 'Genetic Rescue' Means," Revive & Restore, Long Now Foundation, accessed December 6, 2014, http://longnow.org/revive/what-we-do/genetic-rescue/.

"A bit of cloning can" Stewart Brand, "Transcript of 'The Dawn of de-Extinction. Are You Ready?,'" TED, March 2013, accessed December 6, 2014 https://www.ted.com/talks/stewart _brand_the_dawn_of_de_extinction_are_you_ready/transcript.

"It was a wonderful movie" "Frequently Asked Questions," Revive & Restore, Long Now Foundation, accessed December 6, 2014, http://longnow.org/revive/faq/.

"the de-extinction of animals" Antonio Regalado, "De-Extinction Startup, Ark Corporation, Could Engineer Animals, Humans." *MIT Technology Review,* March 19, 2013. http://www.tech nologyreview.com/view/512671/a-stealthy-de-extinction-startup/.

"solve death" Antonio Regalado, "Google's New Company Calico to Try to Cheat Death," *MIT Technology Review,* September 18, 2013, http://www.technologyreview.com/view/519456/goo gle-to-try-to-solve-death-lol/.

"The finality of extinction" Peter Matthiessen, *The Peter Matthiessen Reader,* ed. Mckay Jenkins (New York: Vintage, 2000), 7.

"attenuate, even partially" George M. Church and Ed Regis, *Regenesis: How Synthetic Biology Will Reinvent Nature and Ourselves* (New York: Basic Books, 2014), 140.

"boutique" Ibid., 143.

"We had this meeting" Author interview with Joel Greenberg, July15, 2013.

"If the definition of 'endangered'" Jamie Rappaport Clark, "Politics of De-Extinction" (conference presentation, "De-Extinction: Ethics, Law & Politics," Stanford Law School, California, May 31, 2013), https://www.law.stanford.edu/event/2013/05/31/de-extinction-ethics-law-politics.

"Revived species are cool" Ibid.

"playing God" Jay Odenbaugh, "Hubris and Naturalness" (conference presentation at "De-Extinction: Ethics, Law & Politics," Stanford University, California, May 31, 2013), https://www .law.stanford.edu/event/2013/05/31/de-extinction-ethics-law-politics.

"Define moral hazard" "Justice, Hubris, and Moral Issues." Conference presentation, "De-Extinction: Ethics, Law & Politics, Stanford University, California, May 31, 2013. https://www.law .stanford.edu/event/2013/05/31/de-extinction-ethics-law-politics.

"non-tragic relationship to nature" Jay Odenbaugh. "Justice, Hubris, and Moral Issues." Conference presentation, "De-Extinction: Ethics, Law & Politics," Stanford University, California, May 31, 2013. https://www.law.stanford.edu/event/2013/05/31/de-extinction-ethics-law-politics.

"Shipment of 100 barrels" Schorger, "Great Wisconsin Passenger Pigeon Nesting of 1871," 23.

"until the road was dotted" Schorger, *Passenger Pigeon,* vii.

"Deep, youthful impressions" Ibid.

"The conclusion is inescapable" Ibid., 229.

"This species was in danger" Ibid., 223.

"social facilitation at low densities" Enrique H. Bucher, "The Causes of Extinction of the Passenger Pigeon," *Current Ornithology,* volume 9 (New York: Plenum Press, 1992), 2.

"sucking up the laden fruits" Aldo Leopold, *A Sand County Almanac* (New York: Ballantine Books, 1986), 118.

"As moral agents" Eric Katz, *Nature as Subject* (Lanham, Maryland: Rowman & Littlefield, 1996), xxv.

"interspecific chimeras" Marie-Cecile Van de Lavoir et al., "Interspecific Germline Transmission of Cultured Primordial Germ Cells," *PLoS ONE* 7, no. 5 (May 21, 2012): e35664, doi:10.1371/journal.pone.0035664.

"After two or three years" Ben Novak, "How to Bring Passenger Pigeons All the Way Back," March 15, 2013. Revive & Restore, Long Now Foundation, accessed December 6, 2014, http://longnow.org/revive/events/tedxdeextinction/.

"outbreak population" Charles C. Mann, "Unnatural Abundance," Opinion sec., *New York Times,* November 25, 2004.

CHAPTER 8: NICE TO MEET YOU, NEANDERTHAL

"what did happen to your" L. Sprague de Camp, "The Gnarly Man," *Modern Classics of Fantasy.* edited by Gardner Dozois. (New York: St. Martin's Press, 1997), 26.

"They could maybe even create" "George Church Explains How DNA Will Be Construction Material of the Future," *Spiegel Online,* January 18, 2013, http://www.spiegel.de/international/zeitgeist/george-church-explains-how-dna-will-be-construction-material-of-the-future-a-877634.html.

"the question arises whether" George M. Church and Ed Regis, *Regenesis: How Synthetic Biology Will Reinvent Nature and Ourselves* (New York: Basic Books, 2014), 137.

"Wanted: 'Adventurous woman'" Allan Hall and Fiona Macrae, "Wanted: 'Adventurous Woman' to Give Birth to Neanderthal Man—Harvard Professor Seeks Mother for Cloned Cave Baby," *Daily Mail,* January 20, 2013. http://www.dailymail.co.uk/news/article-2265402/Adventurous-human-woman-wanted-birth-Neanderthal-man-Harvard-professor.html.

"Neanderthals were sentient human beings" Svante Pääbo, "Neanderthals Are People, Too," *New York Times,* April 24, 2014, http://www.nytimes.com/2014/04/25/opinion/neanderthals-are-people-too.html.

"constantly expanding upward range" Stephen Jay Gould, *Wonderful Life: The Burgess Shale and the Nature of History* (New York: W. W. Norton, 1990), 233.

"In direct competition with Cro-Magnons" Thomas Wynn and Frederick L. Coolidge, *How To Think Like a Neandertal* (New York: Oxford University Press, 2011), 187.

"Neanderthal people were" Gould, *Wonderful Life,* 233.

"Why did one type" Pääbo, "Neanderthals Are People, Too."

"lacked reflective awareness" Max Oelschlaeger. *The Idea of Wilderness: From Prehistory to the Age of Ecology.* (New Haven: Yale University Press, 1993), 11.

"The tree-in-itself" Andrew Iliadis, "Interview with Graham Harman (2)," *Figure/Ground: An Open-Source, Para-Academic, Inter-Disciplinary Collaboration,* October 2, 2013, http://figureground.org/interview-with-graham-harman-2/.

"Given that almost every problem" Richard Grusin, ed., *The Nonhuman Turn* (Minneapolis: University of Minnesota Press, 2015), vii.

"An idea, a relationship" Bill McKibben, *The End of Nature* (New York: Random House Trade, 2006), 41.

CODA: *ULTIMA THULE:* ENDS OF THE EARTH

"consciousness of space" Christiane Ritter, *A Woman in the Polar Night* (1938; repr., Fairbanks: University of Alaska Press, 2010), 202.

"told of journeys" Ibid., 12.

"It won't be too lonely" Ibid.

"thick books in the remote quiet" Ibid.

"arid picture of death" Ibid., 30.

"One day melts into" Ibid., 41.

"when the reality of" Ibid., 110.

"After a while my" Ibid., 94.

"The power of this" Ibid., 98.

"as in biblical times" Ibid., 102.

"stronger than all reason" Ibid., 211.

"tiny piece of coal" Ibid., 109.

"unfathomable gulf between" Ibid., 136.

"regional climate models" C. Lang, X. Fettweis, and M. Erpicum. "Stable Climate and Surface Mass Balance in Svalbard over 1979–2013 despite the Arctic Warming." *The Cryosphere* 9, no. 1 (January 8, 2015): 83. doi:10.5194/tc-9-83-2015.

"We see that there" Nilsen Thomas, "No Ice—No Cubs," *Barentsobserver,* June 27, 2012, http://barentsobserver.com/en/nature/no-ice-no-cubs-27-06.

"Every head of wild life" Aldo Leopold. *Game Management.* (Madison, Univ of Wisconsin Press, 1987) xviii.

SELECTED BIBLIOGRAPHY

Adams, Douglas, and Mark Carwardine. *Last Chance to See*. Reprint edition. New York: Ballantine Books, 1992.

Agapow, Paul-Michael, Olaf R. P. Bininda-Emonds, Keith A. Crandall, John L. Gittleman, Georgina M. Mace, Jonathon C. Marshall, and Andy Purvis. "The Impact of Species Concept on Biodiversity Studies." *The Quarterly Review of Biology*, Vol. 79, No. 2, June 2004. doi:10.1086/383542.

Aguilar, A. "A Review of Old Basque Whaling and Its Effect on the Right Whales (Eubalaena Glacialis) of the North Atlantic." *Report of the International Whaling Commission* (Special Issue), Vol. 10, 1986.

Alexander, Helen K., Guillaume Martin, Oliver Y. Martin, and Sebastian Bonhoeffer. "Evolutionary Rescue: Linking Theory for Conservation and Medicine." *Evolutionary Applications*, Vol. 7, Issue 10, December 2014. doi:10.1111/eva.12221.

Allendorf, Fred W., Paul A. Hohenlohe, and Gordon Luikart. "Genomics and the Future of Conservation Genetics." *Nature Reviews Genetics*, Vol. 11, No. 10, October 2010. doi:10.1038/nrg2844.

Allendorf, Fred W., Robb F. Leary, Paul Spruell, and John K. Wenburg. "The Problems with Hybrids: Setting Conservation Guidelines." *Trends in Ecology & Evolution*, Vol.16, No. 11 (2001).

Alvarez, Ken. *Twilight of the Panther: Biology, Bureaucracy and Failure in an Endangered Species Program*. Sarasota: Myakka River Publishing, 1993.

Amato, George D. "Species Hybridization and Protection of Endangered Animals." *Science*, Vol. 253, No. 5017, 1991.

Amato, George, Howard C. Rosenbaum, and Rob DeSalle. *Conservation Genetics in the Age of Genomics*. New York: Columbia University Press, 2009.

Anthes, Emily. *Frankenstein's Cat: Cuddling Up to Biotech's Brave New Beasts*. New York: Scientific American/Farrar, Straus and Giroux, 2014.

Arch, Victoria S., Corinne L. Richards-Zawaki, and Albert S. Feng. "Acoustic Communication in the Kihansi Spray Toad (Nectophrynoides Asperginis): Insights from a Captive Population." *Journal of Herpetology*, Vol. 45, No. 1, March 1, 2011. doi:10.1670/10-084.1.

Askins, Robert A. *Restoring North America's Birds: Lessons from Landscape Ecology*. New Haven: Yale University Press, 2000.

Avant, Deborah D. *The Market for Force: The Consequences of Privatizing Security*. Cambridge, UK, and New York: Cambridge University Press, 2005.

Barkham, Selma Huxley. "The Basque Whaling Establishments in Labrador 1536–1632: A Summary." *Arctic*, Vol. 37, No. 4, December 1984.

Barrow, Mark V. Jr,. *Nature's Ghosts: Confronting Extinction from the Age of Jefferson to the Age of Ecology.* First edition. Chicago and London: University of Chicago Press, 2009.

Bass, Rick. *The Ninemile Wolves.* Boston: Mariner Books, 2003.

Bell, Michael A., and Windsor E. Aguirre. "Contemporary Evolution, Allelic Recycling, and Adaptive Radiation of the Threespine Stickleback." *Evolutionary Ecology Research,* Vol. 15, 2013.

Biermann, Christine, and Becky Mansfield. "Biodiversity, Purity, and Death: Conservation Biology as Biopolitics." *Environment and Planning D: Society and Space,* Vol. 32, No. 2, 2014. doi:10.1068/d13047p.

Blockstein, D. E. "Lyme Disease and the Passenger Pigeon?" *Science,* Vol. 279, No. 5358, March 20, 1998. doi:10.1126/science.279.5358.1831c.

Bogost, Ian. *Alien Phenomenology, or What It's Like to Be a Thing.* Minneapolis: University of Minnesota Press, 2012.

Brand, Stewart. *Whole Earth Discipline: Why Dense Cities, Nuclear Power, Transgenic Crops, Restored Wildlands, and Geoengineering Are Necessary.* New York: Penguin Books, 2010.

Brown, David E., ed. *The Wolf in the Southwest: The Making of an Endangered Species.* Silver City, NM: High Lonesome Books, 2002.

Bruce, Donald, and Ann Bruce. *Engineering Genesis: Ethics of Genetic Engineering in Non-Human Species.* New York: Routledge, 2014.

Bryant, Levi R. *The Democracy of Objects.* Ann Arbor: Open Humanities Press/Michigan Publishing, University of Michigan Library, 2011.

Bucher, Enrique H. "The Causes of Extinction of the Passenger Pigeon." *Current Ornithology,* Vol. 9. Dennis M. Power, ed. New York: Plenum Press, 1992.

Burgess, N. D., T. M. Butynski, N. J. Cordeiro, N. H. Doggart, J. Fjeldså, K. M. Howell, F. B. Kilahama, et al. "The Biological Importance of the Eastern Arc Mountains of Tanzania and Kenya." *Biological Conservation,* Vol. 134, No. 2, January 2007. doi:10.1016/j.biocon.2006.08.015.

Burgess, N. D., J. Fjeldså, and R. Botterweg. "Faunal Importance of the Eastern Arc Mountains of Kenya and Tanzania." *Journal of East African Natural History,* Vol. 87, No. 1, January 1, 1998. doi:10.2982/0012-8317(1998)87[37:FIOTEA]2.0.CO;2.

Callicott, J. Baird. *Beyond the Land Ethic: More Essays in Environmental Philosophy.* Albany: State University of New York Press, 1999.

———. "Rolston on Intrinsic Value: A Deconstruction." *Environmental Ethics,* Vol. 14, No. 2, 1992. doi:10.5840/enviroethics199214229.

Carroll, Scott P., and Charles W. Fox, eds. *Conservation Biology: Evolution in Action.* Oxford, UK, and New York: Oxford University Press, 2008.

Carroll, S. P., P. S. Jorgensen, M. T. Kinnison, C. T. Bergstrom, R. F. Denison, P. Gluckman, T. B. Smith, S. Y. Strauss, and B. E. Tabashnik. "Applying Evolutionary Biology to Address Global Challenges." *Science,* Vol. 346, No. 6207, October 17, 2014. doi:10.1126/science.1245993.

Chernela, Janet. "A Species Apart: Ideology, Science, and the End of Life." In *The Anthropology of Extinction: Essays on Culture and Species Death,* edited by Genese Marie Sodikoff. Bloomington: Indiana University Press, 2012.

Church, George M., and Ed Regis. *Regenesis: How Synthetic Biology Will Reinvent Nature and Ourselves.* New York: Basic Books, 2014.

Cole, Timothy V. N., Philip Hamilton, Allison Glass Henry, Peter Duley, Richard M. Pace, Bradley N. White, and Tim Frasier. "Evidence of a North Atlantic Right Whale Eubalaena Glacialis Mating Ground." *Endangered Species Research,* Vol. 21, No. 1, July 3, 2013. doi:10.3354/esr00507.

Collins, James P., and Andrew Storfer. "Global Amphibian Declines: Sorting the Hypotheses." *Diversity and Distributions,* Vol. 9, No. 2, March 1, 2003. doi:10.1046/j.1472-4642.2003.00012.x.

Collyer, Michael L., Jeffrey S. Heilveil, and Craig A. Stockwell. "Contemporary Evolutionary Divergence for a Protected Species Following Assisted Colonization." *PLoS ONE,* Vol. 6, No. 8, e22310, August 31, 2011. doi:10.1371/journal.pone.0022310.

Collyer, Michael L., James M. Novak, Craig A. Stockwell, and M. E. Douglas. "Morphological Divergence of Native and Recently Established Populations of White Sands Pupfish (Cyprinodon Tularosa)." *Copeia,* No. 1, 2005.

Corthals, Angelique, and Rob Desalle. "An Application of Tissue and DNA Banking for Genomics and Conservation: The Ambrose Monell Cryo-Collection (AMCC)." *Systematic Biology*, Vol. 54, No. 5, October 1, 2005. doi:10.1080/10635150590950353.

Costello, M. J., R. M. May, and N. E. Stork. "Can We Name Earth's Species Before They Go Extinct?" *Science*, Vol. 339, No. 6118, January 25, 2013. doi:10.1126/science.1230318.

Craig Pittman. "Saga of Florida Panther Is 'Sordid Story.'" *Tampa Bay Times*, April 16, 2010. http://www.tampabay.com/news/environment/wildlife/saga-of-florida-panther-is-sordid-story/1087965.

Darwin, Charles. *On the Origin of Species by Means of Natural Selection, or the Preservation of Favoured Races in the Struggle for Life*. London: W. Clowes and Sons, 1859.

Day, J. J., J. L. Bamber, P. J. Valdes, and J. Kohler. "The Impact of a Seasonally Ice Free Arctic Ocean on the Temperature, Precipitation and Surface Mass Balance of Svalbard." *The Cryosphere*, Vol. 6, No. 1, January 10, 2012. doi:10.5194/tc-6-35-2012.

Delord, Julien. "Can We Really Re-Create an Extinct Species by Cloning?" In *The Ethics of Animal Re-Creation and Modification: Reviving, Rewilding, Restoring*, edited by Markku Oksanen and Helena Siipi. New York: Palgrave Macmillan, 2014.

DeSalle, Rob, and George Amato. "The Expansion of Conservation Genetics." *Nature Reviews Genetics*, Vol. 5, No. 9, September 2004. doi:10.1038/nrg1425.

Dolin, Eric Jay. *Leviathan: The History of Whaling in America*. New York: W. W. Norton & Company, 2008.

Eldredge, Niles. *Reinventing Darwin: Great Evolutionary Debate*. London: Weidenfeld & Nicolson, 1995.

Elliot, Robert. "Faking Nature." *Inquiry*, Vol. 25, No. 1, January 1, 1982. doi:10.1080/00201748208601955.

Fiege, Mark. *The Republic of Nature: An Environmental History of the United States*. Reprint edition. Seattle: University of Washington Press, 2013.

Fisher, Diana O., and Simon P. Blomberg. "Correlates of Rediscovery and the Detectability of Extinction in Mammals." *Proceedings of the Royal Society of London B: Biological Sciences*, Vol. 278, No. 1708 (April 7, 2011). doi:10.1098/rspb.2010.1579.

Fisher, Matthew C., and Trenton W. J. Garner. "The Relationship between the Emergence of Batrachochytrium Dendrobatidis, the International Trade in Amphibians and Introduced Amphibian Species." *Fungal Biology Reviews*, Vol. 21, No. 1, February 2007. doi:10.1016/j.fbr.2007.02.002.

Fitch, W. M., and F. J. Ayala. "Tempo and Mode in Evolution." *Proceedings of the National Academy of Sciences of the United States of America*, Vol. 91, No. 15, July 19, 1994.

Fletcher, Amy L. "Mendel's Ark: Conservation Genetics and the Future of Extinction." *Review of Policy Research*, Vol. 25, No. 6, 2008. doi:10.1111/j.1541-1338.2008.00367_1.x.

Folch, J., M. J. Cocero, P. Chesné, J. L. Alabart, V. Domínguez, Y. Cognié, A. Roche, et al. "First Birth of an Animal from an Extinct Subspecies (Capra Pyrenaica Pyrenaica) by Cloning." *Theriogenology*, Vol. 71, No. 6, April 1, 2009. doi:10.1016/j.theriogenology.2008.11.005.

Frankel, Otto H. "Genetic Conservation: Our Evolutionary Responsibility." *Genetics*, Vol. 78, No. 1, 1974.

Frankham, Richard, Jonathan D. Ballou, and David A. Briscoe. *Introduction to Conservation Genetics*. 2nd edition. Cambridge: Cambridge University Press, 2010.

Frankham, Richard, Jonathan D. Ballou, Michele R. Dudash, Mark D. B. Eldridge, Charles B. Fenster, Robert C. Lacy, Joseph R. Mendelson, Ingrid J. Porton, Katherine Ralls, and Oliver A. Ryder. "Implications of Different Species Concepts for Conserving Biodiversity." *Biological Conservation*, Vol. 153, September 2012. doi:10.1016/j.biocon.2012.04.034.

Franklin, I. R., and R. Frankham. "How Large Must Populations Be to Retain Evolutionary Potential?" *Animal Conservation*, Vol. 1, No. 1, February 1998. doi:10.1017/S136794309821103.

Frasier, T. R., P. K. Hamilton, M. W. Brown, L. A. Conger, A. R. Knowlton, M. K. Marx, C. K. Slay, S. D. Kraus, and B. N. White. "Patterns of Male Reproductive Success in a Highly Promiscuous Whale Species: The Endangered North Atlantic Right Whale." *Molecular Ecology*, Vol. 16, No. 24, December 2007. doi:10.1111/j.1365-294X.2007.03570.x.

Friedrich Ben-Nun, Inbar, Susanne C. Montague, Marlys L. Houck, Ha T. Tran, Ibon Gari-taonandia, Trevor R. Leonardo, Yu-Chieh Wang, et al. "Induced Pluripotent Stem Cells from Highly Endangered Species." *Nature Methods,* Vol. 8, No. 10. September 4, 2011. doi:10.1038/nmeth.1706.

Fujiwara, Masami, and Hal Caswell. "Demography of the Endangered North Atlantic Right Whale." *Nature,* Vol. 414, No. 6863, November 29, 2001. doi:10.1038/35107054.

Genome 10K Community of Scientists. "Genome 10K: A Proposal to Obtain Whole-Genome Sequence for 10,000 Vertebrate Species." *Journal of Heredity,* Vol. 100, No. 6, November 1, 2009. doi:10.1093/jhered/esp086.

Ghiselin, Michael T. "A Radical Solution to the Species Problem." *Systematic Biology,* Vol. 23, No. 4, December 1, 1974. doi:10.1093/sysbio/23.4.536.

Gingerich, P. D. "Quantification and Comparison of Evolutionary Rates." *American Journal of Science,* Vol. 293-A, January 1, 1993. doi:10.2475/ajs.293.A.453.

Gonzalez, Andrew, Ophélie Ronce, Regis Ferriere, and Michael E. Hochberg. "Evolutionary Rescue: An Emerging Focus at the Intersection between Ecology and Evolution." *Philosophical Transactions of the Royal Society B: Biological Sciences,* Vol. 368, No. 1610, January 19, 2013. doi:10.1098/rstb.2012.0404.

Gould, Stephen Jay. *An Urchin in the Storm: Essays about Books and Ideas.* New York: W. W. Norton & Company, 1988.

———. *Wonderful Life: The Burgess Shale and the Nature of History.* New York: W. W. Norton & Company, 1990.

———. "Tempo and Mode in the Macroevolutionary Reconstruction of Darwinism." *Proceedings of the National Academy of Sciences of the United States of America,* Vol. 91, No. 15, July 19, 1994.

Greenberg, Joel. *A Feathered River Across the Sky: The Passenger Pigeon's Flight to Extinction.* New York: Bloomsbury, 2014.

Greene, Charles H., Andrew J. Pershing, Robert D. Kenney, and Jack W. Jossi. "Impact of Climate Variability on the Recovery of Endangered North Atlantic Right Whales." *Oceanography,* Vol 16, No. 4, 2003.

Grenier, Robert. "The Basque Whaling Ship from Red Bay, Labrador: A Treasure Trove of Data on Iberian Atlantic Shipbuilding Design and Techniques in the Mid-16th Century." In *Trabalhos de Arqueologia 18—Proceedings. International Symposium on Archaeology of Medieval and Modern Ships of Iberian-Atlantic Tradition. Hull Remains, Manuscripts and Ethnographic Sources: A Comparative Approach,* ed. Francisco Alves. Lisbon: Centro Nacional de Arqueologia Nautica e Subaquatica/Academia de Marinha, 1998.

Grusin, Richard, ed. *The Nonhuman Turn.* Minneapolis: University of Minnesota Press, 2015.

Halliday, T. R. "The Extinction of the Passenger Pigeon Ectopistes Migratorius and Its Relevance to Contemporary Conservation." *Biological Conservation,* Vol. 17, 1980.

Hambler, Clive, Peter A. Henderson, and Martin R. Speight. "Extinction Rates, Extinction-Prone Habitats, and Indicator Groups in Britain and at Larger Scales." *Biological Conservation,* Vol. 144, No. 2 (February 2011). doi:10.1016/j.biocon.2010.09.004.

Harman, Graham. *Guerrilla Metaphysics: Phenomenology and the Carpentry of Things.* Chicago: Open Court, 2005.

Harrison, K. David. *When Languages Die: The Extinction of the World's Languages and the Erosion of Human Knowledge.* Oxford: Oxford University Press, 2008.

Heatherington, Tracey. "From Ecocide to Genetic Rescue: Can Technoscience Save the Wild?" In *The Anthropology of Extinction: Essays on Culture and Species Death,* edited by Genese Marie Sodikoff. Bloomington and Indianapolis: Indiana University Press, 2012.

Hedrick, Philip W. "Gene Flow and Genetic Restoration: The Florida Panther as a Case Study." *Conservation Biology,* Vol, 9, No. 5, October 1, 1995. doi:10.1046/j.1523-1739.1995.9050988.x-i1.

Hedrick, Philip W., and Richard Fredrickson. "Genetic Rescue Guidelines with Examples from Mexican Wolves and Florida Panthers." *Conservation Genetics,* Vol. 11, No. 2, April 2010. doi:10.1007/s10592-009-9999-5.

Hedrick, P. W., and R. J. Fredrickson. "Captive Breeding and the Reintroduction of Mexican and Red Wolves." *Molecular Ecology*, Vol. 17, No. 1, January 2008. doi:10.1111/j.1365-294X.2007.03400.x.

He, Fangliang, and Stephen P. Hubbell. "Species-Area Relationships Always Overestimate Extinction Rates from Habitat Loss." *Nature*, Vol. 473, No. 7347, May 19, 2011. doi:10.1038/nature09985.

Hendry, A. P., and M. T. Kinnison. "An Introduction to Microevolution: Rate, Pattern, Process." *Genetica*, Vol. 112–113, November 1, 2001. doi:10.1023/A:1013368628607.

———. "The Pace of Modern Life: Measuring Rates of Contemporary Microevolution." *Evolution*, Vol. 53, No. 6, December 1999.

———. "The Pace of Modern Life II: From Rates of Contemporary Microevolution to Pattern and Process." *Genetica*, Vol. 112–113, 2001. doi:10.1023/A:1013375419520

Hey, Jody. *Genes, Categories, and Species: The Evolutionary and Cognitive Cause of the Species Problem.* Oxford: Oxford University Press, 2001.

Hickey, Joseph J. "In Memoriam: Arlie William Schorger." *The Auk,* Vol. 90, July 1973.

Hillman Smith, Kes, and Fraser Smith. "Conservation Crises and Potential Solutions: Example of Garamba National Park Democratic Republic of Congo." Presented at the Second World Congress of the International Ranger Federation. Costa Rica, September 25, 1997.

Holmberg, Tora, Nete Schwennesen, and Andrew Webster. "Bio-Objects and the Bio-Objectification Process." *Croatian Medical Journal*, Vol. 52, No. 6 (December 2011): 740–42. doi:10.3325/cmj.2011.52.740.

Hostetler, Jeffrey A., David P. Onorato, Deborah Jansen, and Madan K. Oli. "A Cat's Tale: The Impact of Genetic Restoration on Florida Panther Population Dynamics and Persistence." *Journal of Animal Ecology,* Vol. 82, No. 3, May 2013. doi:10.1111/1365-2656.12033.

Hunter Clark, ed. *The Life and Letters of Alexander Wilson.* Vol. 154. Philadelphia: Memoirs of the American Philosophical Society 1983.

Iliadis, Andrew. "Interview with Graham Harman (2)." *Figure/Ground: An Open-Source, Para-Academic, Inter-Disciplinary Collaboration,* October 2, 2013. http://figureground.org/interview-with-graham-harman-2/.

Johnson, Phillip. "The Extinction of Darwinism: Review of 'Extinction: Bad Gene or Bad Luck' by David M. Raup." *The Atlantic,* February 1992. http://www.arn.org/docs/johnson/raup.htm.

Johnson, W. E., D. P. Onorato, M. E. Roelke, E. D. Land, M. Cunningham, R. C. Belden, R. McBride, et al. "Genetic Restoration of the Florida Panther." *Science,* Vol. 329, No. 5999, September 24, 2010. doi:10.1126/science.1192891.

Kaliszewska, Zofia A., Jon Seger, Victoria J. Rowntree, Susan G. Barco, Rafael Benegas, Peter B. Best, Moira W. Brown, et al. "Population Histories of Right Whales (Cetacea: Eubalaena) Inferred from Mitochondrial Sequence Diversities and Divergences of Their Whale Lice (Amphipoda: Cyamus)." *Molecular Ecology,* Vol. 14, No. 11, October 2005. doi:10.1111/j.1365-294X.2005.02664.x.

Kaplan, Matt. *The Science of Monsters: The Origins of the Creatures We Love to Fear.* New York: Simon and Schuster, 2013.

Katz, Eric. *Nature as Subject.* Lanham: Rowman & Littlefield Publishers, 1996.

Katz, Eric, and Andrew Light, eds. *Environmental Pragmatism.* London: Routledge, 1996.

Kautz, Randy, Robert Kawula, Thomas Hoctor, Jane Comiskey, Deborah Jansen, Dawn Jennings, John Kasbohm, et al. "How Much Is Enough? Landscape-Scale Conservation for the Florida Panther." *Biological Conservation,* Vol. 130, No. 1, June 2006. doi:10.1016/j.biocon.2005.12.007.

Kosek, Jake. *Understories: The Political Life of Forests in Northern New Mexico.* Durham: Duke University Press Books, 2006.

Kouba, Andrew J., Rhiannon E. Lloyd, Marlys L. Houck, Aimee J. Silla, Natalie Calatayud, Vance L. Trudeau, John Clulow, et al. "Emerging Trends for Biobanking Amphibian Genetic Resources: The Hope, Reality and Challenges for the Next Decade." *Biological Conservation,* Vol. 164, August 2013. doi:10.1016/j.biocon.2013.03.010.

Kraus, Scott D., and Rosalind M. Rolland. *The Urban Whale: North Atlantic Right Whales at the Crossroads,* Cambridge, MA: Harvard University Press, 2010.

Krisch, Joshua A. "New Study Offers Clues to Swift Arctic Extinction." *The New York Times,* August 28, 2014.

Lang, C., X. Fettweis, and M. Erpicum. "Stable Climate and Surface Mass Balance in Svalbard over 1979–2013 despite the Arctic Warming." *The Cryosphere,* Vol. 9, No. 1 (January 8, 2015): 83–101. doi:10.5194/tc-9-83-2015.

Lee, S., K. Zippel, L. Ramos, and J. Searle. "Captive-Breeding Programme for the Kihansi Spray Toad Nectophrynoides Asperginis at the Wildlife Conservation Society, Bronx, New York." *International Zoo Yearbook,* Vol. 40, No. 1, July 1, 2006. doi:10.1111/j.1748-1090.2006.00241.x.

Leopold, Aldo. *A Sand County Almanac.* New York: Ballantine Books, 1986.

———. *Game Management.* Madison: University of Wisconsin Press, 1987.

Lestel, Dominique. "The Withering of Shared Life through the Loss of Biodiversity." *Social Science Information,* Vol. 52, No. 2, June 1, 2013. doi:10.1177/0539018413478325.

Lévi-Strauss, Claude. *The Savage Mind.* Chicago: University of Chicago Press, 1966.

Levy, Sharon. *Once and Future Giants: What Ice Age Extinctions Tell Us About the Fate of Earth's Largest Animals.* Oxford: Oxford University Press, 2011.

Lieberman, Alan. "Alala Egg That Changed the Future." Hawaiian Birds, San Diego Zoo, January 8, 2013. http://blogs.sandiegozoo.org/2013/01/08/alala-egg-changed-future/.

Light, Andrew, and Holmes Rolston III, eds. *Environmental Ethics: An Anthology.* Malden: Wiley-Blackwell, 2002.

Lippsett, Lonny. "Diving into the Right Whale Gene Pool." *Oceanus Magazine,* Vol. 44, No. 3, December 3, 2005.

Lopez, Barry. *Arctic Dreams.* New York: Vintage, 2001.

MacPhee, R. D. E. *Extinctions in Near Time.* New York: Springer Science & Business Media, 1999.

Maehr, David. *The Florida Panther: Life and Death of a Vanishing Carnivore.* Washington, DC: Island Press, 1997.

Maehr, D. S., P. Crowley, J. J. Cox, M. J. Lacki, J. L. Larkin, T. S. Hoctor, L. D. Harris, and P. M. Hall. "Of Cats and Haruspices: Genetic Intervention in the Florida Panther. Response to Pimm et al. (2006)." *Animal Conservation,* Vol. 9, No. 2, May 2006. doi:10.1111/j.1469-1795. 2005.00019.x.

Mann, Charles C. "Unnatural Abundance." *The New York Times,* November 25, 2004, opinion section.

Marchant, Jo. "Evolution Machine: Genetic Engineering on Fast Forward." *New Scientist,* Issue 2818, June 27, 2011.

Martinelli, Lucia, Markku Oksanen, and Helena Siipi. "De-Extinction: A Novel and Remarkable Case of Bio-Objectification." *Croatian Medical Journal,* Vol. 55, No. 4, August 2014. doi:10.3325/cmj.2014.55.423.

Martínez-Moreno, Jorge, Rafael Mora, and Ignacio de la Torre. "The Middle-to-Upper Palaeolithic Transition in Cova Gran (Catalunya, Spain) and the Extinction of Neanderthals in the Iberian Peninsula." *Journal of Human Evolution,* Vol. 58, No. 3, March 2010. doi:10.1016/j. jhevol.2009.09.002.

Marzluff, John M., Tony Angell, and Paul R. Ehrlich. *In the Company of Crows and Ravens.* New Haven: Yale University Press, 2007.

Matthiessen, Peter. *Wildlife in America.* New York: Penguin Books, 1978.

———. *The Peter Matthiessen Reader.* Edited by McKay Jenkins. New York: Vintage, 2000.

———. *The Snow Leopard.* New York: Penguin Classics, 2008.

———. *African Silences.* New York: Vintage, 2012.

Matthiessen, Peter, and Maurice Hornocker. *Tigers in the Snow.* New York: North Point Press, 2001.

Mayr, Ernst. "What Is a Species, and What Is Not?" *Philosophy of Science.* Vol. 63, No. 2, June 1996.

McBride, Roy T. *The Mexican Wolf (Canis Lupus Baileyi): A Historical Review and Observations on Its Status and Distribution: A Progress Report to the U.S. Fish and Wildlife Service.* U.S. Fish and Wildlife Service, 1980.

———. "Three Decades of Searching South Florida for Panthers." Presented at the Proceedings of The Florida Panther Conference, Fort Myers, Florida, November 1, 1994.

McCabe, Robert A. "A. W. Schorger: Naturalist and Writer." *The Passenger Pigeon*, Vol. 55, No. 4, Winter 1993.

McCarthy, Cormac. *The Crossing*. New York: Alfred A. Knopf, 1994.

McCarthy, Michael A., Colin J. Thompson, and Stephen T. Garnett. "Optimal Investment in Conservation of Species." *Journal of Applied Ecology*. Vol. 45, No. 5, October 1, 2008. doi:10.1111/j.1365-2664.2008.01521.x.

McKibben, Bill. *The End of Nature*. New York: Random House Trade Paperbacks, 2006.

McLeod, B. A., Moira W. Brown, Michael J. Moore, W. Stevens, Selma H. Barkham, Michael Barkham, and B. N. White. "Bowhead Whales, and Not Right Whales, Were the Primary Target of 16th-to 17th-Century Basque Whalers in the Western North Atlantic." *Arctic*, Vol. 61, No. 1, 2008.

McLeod, Brenna A., Moira W. Brown, Timothy R. Frasier, and Bradley N. White. "DNA Profile of a Sixteenth Century Western North Atlantic Right Whale (Eubalaena Glacialis)." *Conservation Genetics*, Vol. 11, No. 1, February 2010. doi:10.1007/s10592-009-9811-6.

Meinzer, Oscar Edward, and Raleigh Frederick Hare. *Geology and Water Resources of Tularosa Basin, New Mexico*, Vol. 343. Washington, DC: United States Geological Survey, Department of the Interior, 1915.

Melville, Herman. *Moby Dick: Or the Whale*. London: Modern Library, 1992.

Milledge, Simon A. H. "Illegal Killing of African Rhinos and Horn Trade, 2000–2005: The Era of Resurgent Markets and Emerging Organized Crime." *Pachyderm*, No. 43, 2007.

Miller, Claire. "Great Barrier Reef 'on Ice.'" *Frontiers in Ecology and the Environment*, Vol.10, No. 2, March 2012.

Miller, Robert Rush, and Anthony A. Echelle. "Cyprinodon Tularosa, a New Cyprinodontid Fish from the Tularosa Basin, New Mexico." *The Southwestern Naturalist*. Vol. 19, No. 4, January 20, 1975. doi:10.2307/3670395.

Miller, Webb, Vanessa M. Hayes, Aakrosh Ratan, Desiree C. Petersen, Nicola E. Wittekindt, Jason Miller, Brian Walenz, et al. "Genetic Diversity and Population Structure of the Endangered Marsupial Sarcophilus Harrisii (Tasmanian Devil)." *Proceedings of the National Academy of Sciences*, Vol. 108, No. 30, July 26, 2011. doi:10.1073/pnas.1102838108.

Milot, E., H. Weimerskirch, P. Duchesne, and L. Bernatchez. "Surviving with Low Genetic Diversity: The Case of Albatrosses." *Proceedings of the Royal Society B: Biological Sciences*, Vol. 274, No. 1611, March 22, 2007. doi:10.1098/rspb.2006.0221.

Minard, Anne. "West Nile Devastated Bird Species." *National Geographic News*, May 16, 2007.

———. "'Reverse Evolution' Discovered in Seattle Fish." *National Geographic News*, May 20, 2008.

Moore, Michael J. "Rosita Voyage Log." *"Rosita"—Voyage of Discovery*, 2004. www.whale.wheelock.edu/Rosita/.

Morton, Timothy. *Ecology without Nature: Rethinking Environmental Aesthetics*. Cambridge: Harvard University Press, 2009.

———. "Here Comes Everything: The Promise of Object-Oriented Ontology." *Qui Parle: Critical Humanities and Social Sciences*, Vol. 19, No. 2, 2011.

———. "Sublime Objects." *Speculations*, Vol. 2, 2011.

———. *Hyperobjects: Philosophy and Ecology after the End of the World*. Minneapolis: University of Minnesota Press, 2013.

Muir, John, and Peter Jenkins. *A Thousand-Mile Walk to the Gulf*. Boston: Mariner Books, 1998.

"Multiplex Automated Genomic Engineering (MAGE): A Machine That Speeds up Evolution Is Revolutionizing Genome Design." *Wyss Institute*. www.wyss.harvard.edu/viewpage/330/.

Myers, Norman. *The Sinking Ark: A New Look at the Problem of Disappearing Species*. Oxford: Pergamon Press, 1979.

Nagel, Thomas. "What Is It Like to Be a Bat?" *The Philosophical Review*, Vol. 83, No. 4, October 1, 1974. doi:10.2307/2183914.

National Resource Council. *The Scientific Bases for the Preservation of the Hawaiian Crow*, 1992. http://www.nap.edu/catalog/2023/the-scientific-bases-for-the-preservation-of-the-hawaiian-crow.

Nelson, Barney, ed. *God's Country or Devil's Playground: The Best Nature Writing from the Big Bend of Texas*. Austin: University of Texas Press, 2002.

Neumann, Thomas W. "Human-Wildlife Competition and the Passenger Pigeon: Population Growth from System Destabilization." *Human Ecology*, Vol. 13, No. 4, December 1985. doi:10.1007/BF01531152.

Newmark, W. D. "Forest Area, Fragmentation, and Loss in the Eastern Arc Mountains: Implications for the Conservation of Biological Diversity." *Journal of East African Natural History*, Vol. 87, No. 1, January 1, 1998. doi:10.2982/0012-8317(1998)87[29:FAFALI]2.0.CO;2.

———. *Conserving Biodiversity in East African Forests: A Study of the Eastern Arc Mountains*. New York: Springer Science & Business Media, 2002.

Norton, Bryan G. "Environmental Ethics and Weak Anthropocentrism." *Environmental Ethics*, Vol. 6, No. 2, 1984. doi:10.5840/enviroethics19846233.

———. *Why Preserve Natural Variety?*. Princeton: Princeton University Press, 1990.

———. "Epistemology and Environmental Values." *Monist*, Vol. 75, No. 2, April 1992.

———. "Why I am Not a Nonanthropocentrist: Callicott and the Failure of Monistic Inherentism." *Environmental Ethics*, Vol. 17, No. 4, 1995. doi:10.5840/enviroethics19951743.

Norton, Bryan G., Michael Hutchins, Elizabeth F. Stevens, and Terry L. Maple, eds. *Ethics on the Ark*. Washington, DC: Smithsonian Books, 1996.

Novak, Ben. "Flights of Fancy: A Tiny Tube of Clear Liquid." Project Passenger Pigeon—Memoirs, Stories, Paintings, Poems. http://passengerpigeon.org/flights.html.

———. "How to Bring Passenger Pigeons All the Way Back." Presentation at the TedX DeExtinction, Washington, DC, March 15, 2013.

O'Brien, Stephen J. *Tears of the Cheetah: The Genetic Secrets of Our Animal Ancestors*. New York: St. Martin's Griffin, 2005.

O'Brien, Stephen J., and Ernst Mayr. "Bureaucratic Mischief: Recognizing Endangered Species and Subspecies." *Science*, Vol. 51, No. 4998, March 8, 1991.

Oelschlaeger, Max. *The Idea of Wilderness: From Prehistory to the Age of Ecology*. New Haven: Yale University Press, 1993.

Oksanen, Markku, and Helena Siipi, eds. *The Ethics of Animal Re-Creation and Modification: Reviving, Rewilding, Restoring*. New York: Palgrave Macmillan, 2014.

Ożgo, Małgorzata. "Rapid Evolution and the Potential for Evolutionary Rescue in Land Snails." *Journal of Molluscan Studies*, May 5, 2014. doi:10.1093/mollus/eyu029.

Palkovacs, Eric P., Michael T. Kinnison, Cristian Correa, Christopher M. Dalton, and Andrew P. Hendry. "Fates beyond Traits: Ecological Consequences of Human-Induced Trait Change." *Evolutionary Applications*, Vol. 5, Mo. 2, February 2012. doi:10.1111/j.1752-4571.2011.00212.x.

Palumbi, Stephen R. *The Evolution Explosion: How Humans Cause Rapid Evolutionary Change*. New York: W. W. Norton & Company, 2002.

Parry, Bronwyn. "The Fate of the Collections: Social Justice and the Annexation of Plant Genetic Resources." In *People, Plants, and Justice: The Politics of Nature Conservation*, ed. Charles Zerner. New York: Columbia University Press, 2000.

Patenaude, N. J., V. A. Portway, C. M. Schaeff, J. L. Bannister, P. B. Best, R. S. Payne, V. J. Rowntree, M. Rivarola, and C. S. Baker. "Mitochondrial DNA Diversity and Population Structure among Southern Right Whales (Eubalaena Australis)." *Journal of Heredity*, Vol. 98, No. 2, January 6, 2007. doi:10.1093/jhered/esm005.

Pershing, Andrew J. and Charles H. Greene. "Climate and the Conservation Biology of North Atlantic Right Whales: Being a Right Whale at the Wrong Time?" Accessed December 4, 2014. http://oceandata.gmri.org/environmentalprediction/docs/FrontiersinEcologyand theEnvironment_2_29-34.pdf.

Pigliucci, Massimo. "Wittgenstein Solves (Posthumously) the Species Problem." *Philosophy Now*, No. 51, March/April, 2005.

Pimm, S. L., L. Dollar, and O. L. Bass. "The Genetic Rescue of the Florida Panther." *Animal Conservation*, Vol. 9, No. 2, May 2006. doi:10.1111/j.1469-1795.2005.00010.x.

Pittenger, John S., and Craig L. Springer. "Native Range and Conservation of the White Sands Pupfish (Cyprinodon Tularosa)." *The Southwestern Naturalist*, Vol. 44, No. 2, June 1999.

Player, Ian, and Alan Paton. *The White Rhino Saga.* New York: Stein and Day, 1973.

Pond, David W., and Geraint A. Tarling. "Phase Transitions of Wax Esters Adjust Buoyancy in Diapausing Calanoides Acutus." *Limnology and Oceanography,* Vol. 56, No. 4, 2011. doi:10.4319/lo.2011.56.4.1310.

Pounds, J. Alan. "Climate and Amphibian Declines." *Nature,* Vol. 410, No. 6829, April 5, 2001. doi:10.1038/35070683.

Pounds, J. Alan, Martín R. Bustamante, Luis A. Coloma, Jamie A. Consuegra, Michael P. L. Fogden, Pru N. Foster, Enrique La Marca, et al. "Widespread Amphibian Extinctions from Epidemic Disease Driven by Global Warming." *Nature,* Vol. 439, No. 7073, January 12, 2006. doi:10.1038/nature04246.

Powell, Alvin. *The Race to Save the World's Rarest Bird: The Discovery and Death of the Po'ouli.* Mechanicsburg, PA: Stackpole Books, 2008.

Poynton, John C., Kim M. Howell, Barry T. Clarke, and Jon C. Lovett. "A Critically Endangered New Species of Nectophrynoides (Anura: Bufonidae) from the Kihansi Gorge, Udzungwa Mountains, Tanzania." *African Journal of Herpetology,* Vol. 47, No. 2, January 1, 1998. doi: 10.1080/21564574.1998.9650003.

Pratt, Thane K., Carter T. Atkinson, Paul Christian Banko, James D. Jacobi, and Bethany Lee Woodworth, eds. *Conservation Biology of Hawaiian Forest Birds: Implications for Island Avifauna.* New Haven: Yale University Press, 2009.

Preston, Christopher J., and Wayne Ouderkirk, eds. *Nature, Value, Duty: Life on Earth with Holmes Rolston, III.* The International Library of Environmental, Agricultural and Food Ethics. Houten: Springer Netherlands, 2010.

Preston, Douglas J. *Dinosaurs in the Attic: An Excursion into the American Museum of Natural History.* New York: St. Martin's Griffin, 1993.

Proença, Vânia, and Henrique Miguel Pereira. "Comparing Extinction Rates: Past, Present, and Future." In *Encyclopedia of Biodiversity.* Elsevier, 2013.

Quammen, David. *The Song of the Dodo: Island Biogeography in an Age of Extinction.* New York: Scribner, 1997.

Radin, J. "Latent Life: Concepts and Practices of Human Tissue Preservation in the International Biological Program." *Social Studies of Science,* Vol. 43, No. 4, August 1, 2013. doi:10.1177/0306312713476131.

Rastogi, Toolika, Moira W. Brown, Brenna A. McLeod, Timothy R. Frasier, Robert Grenier, Stephen L. Cumbaa, Jeya Nadarajah, and Bradley N. White. "Genetic Analysis of 16th-Century Whale Bones Prompts a Revision of the Impact of Basque Whaling on Right and Bowhead Whales in the Western North Atlantic." *Canadian Journal of Zoology,* Vol. 82, No. 10, October 2004. doi:10.1139/z04-146.

Redford, Kent H., George Amato, Jonathan Baillie, Pablo Beldomenico, Elizabeth L. Bennett, Nancy Clum, Robert Cook, et al. "What Does It Mean to Successfully Conserve a (Vertebrate) Species?" *BioScience,* Vol. 61, No. 1, January 2011. doi:10.1525/bio.2011.61.1.9.

Reed, D. H. "Albatrosses, Eagles and Newts, Oh My!: Exceptions to the Prevailing Paradigm Concerning Genetic Diversity and Population Viability?: Genetic Diversity and Extinction." *Animal Conservation,* Vol. 13, No. 5, June 1, 2010. doi:10.1111/j.1469-1795.2010.00353.x.

Regalado, Antonio. "De-Extinction Startup, Ark Corporation, Could Engineer Animals, Humans." *MIT Technology Review,* March 19, 2013. http://www.technologyreview.com/view/512671/a-stealthy-de-extinction-startup/.

———. "Google's New Company Calico to Try to Cheat Death." *MIT Technology Review,* September 18, 2013. http://www.technologyreview.com/view/519456/google-to-try-to-solve-death-lol/.

Revised Recovery Plan for the 'Alala (Corvus Hawaiiensis). Portland, Oregon: U.S. Fish and Wildlife Service, January 27, 2009.

Rexer, Lyle, Rachel Klein, Edward O. Wilson, and American Museum of Natural History. *American Museum of Natural History: 125 Years of Expedition and Discovery.* New York: Harry N. Abrams, 1995.

Reygondeau, Gabriel, and Grégory Beaugrand. "Future Climate-Driven Shifts in Distribution of Calanus Finmarchicus." *Global Change Biology,* Vol. 17, No. 2, February 2011. doi:10.1111/j.1365-2486.2010.02310.x.

Reznick, David A., Heather Bryga, and John A. Endler. "Experimentally Induced Life-History Evolution in a Natural Population." *Nature,* Vol. 346, No. 6282, 1990.

Rice, Kevin J., and Nancy C. Emery. "Managing Microevolution: Restoration in the Face of Global Change." *Frontiers in Ecology and the Environment,* Vol. 1, No. 9, November 2003. doi:10.2307/3868114.

Ridley, Matt. "Counting Species Out." www.rationaloptimist.com, August 27, 2011.

Ritter, Christiane. *A Woman in the Polar Night.* Fairbanks: University of Alaska Press, 2010.

Robert, Jason Scott, and Françoise Baylis. "Crossing Species Boundaries." *American Journal of Bioethics,* Vol, 3, No. 3, 2003.

Rödder, D., J. Kielgast, and S. Lötters. "Future Potential Distribution of the Emerging Amphibian Chytrid Fungus under Anthropogenic Climate Change." *Diseases of Aquatic Organisms,* Vol. 92, No. 3, April 7, 2010. doi:10.3354/dao02197.

Rolston III, Holmes. *Environmental Ethics: Duties to and Values in the Natural World.* Philadelphia: Temple University Press, 1989.

———. "Value in Nature and the Nature of Value." In *Philosophy and the Natural Environment,* edited by Robin Attfield and Andrew Belsey. Royal Institute of Philosophy Supplement, Vol. 36. Cambridge: Cambridge University Press, 1994.

———. *Genes, Genesis, and God: Values and Their Origins in Natural and Human History.* Cambridge: Cambridge University Press, 1999.

———. "What Is a Gene? From Molecules to Metaphysics." *Theoretical Medicine and Bioethics.* Vol. 27, No. 6, December 2006. doi:10.1007/s11017-006-9022-9.

Romer, Paul. "For Richer, for Poorer." *Prospect Magazine: The Leading Magazine of Ideas,* February 2010. http://www.prospectmagazine.co.uk/features/for-richer-for-poorer.

Root, Alan. *Ivory, Apes & Peacocks: Animals, Adventure and Discovery in the Wild Places of Africa.* London: Chatto & Windus, 2012.

Rosen, Rebecca J. "The Climate Is Set to Change 'Orders of Magnitude' Faster Than at Any Other Time in the Past 65 Million Years." *The Atlantic,* August 2, 2013.

Ryder, O. A. "DNA Banks for Endangered Animal Species." *Science,* Vol. 288, No. 5464, April 14, 2000. doi:10.1126/science.288.5464.275.

Sagoff, Mark. "On Preserving the Natural Environment." *Yale Law Journal,* Vol. 84, No. 2, December 1974.

Schaeff, Catherine M., Scott D. Kraus, Moira W. Brown, and Bradley N. White. "Assessment of the Population Structure of Western North Atlantic Right Whales (Eubalaena Glacialis) Based on Sighting and mtDNA Data." *Canadian Journal of Zoology,* Vol. 71, No. 2, February 1, 1993. doi:10.1139/z93-047.

Schorger, A. W. *The Chemistry of Cellulose and Wood.* New York: McGraw-Hill, 1926.

———. "The Great Wisconsin Passenger Pigeon Nesting of 1871." *The Passenger Pigeon: Monthly Bulletin of the Wisconsin Society of Ornithology,* Vol. 1, No. 1, February 1939.

———. *The Passenger Pigeon: Its Natural History and Extinction.* Madison: University of Wisconsin Press, 1955.

Schueler, Donald G. *Incident at Eagle Ranch: Predators as Prey in the American West.* Tucson: University of Arizona Press, 1991.

Seddon, Philip J., Axel Moehrenschlager, and John Ewen. "Reintroducing Resurrected Species: Selecting DeExtinction Candidates." *Trends in Ecology & Evolution,* Vol. 29, No. 3, March 2014. doi:10.1016/j.tree.2014.01.007.

Seto, Sonia J. "North Atlantic Right Whale DNA." *Right Whale News,* Vol. 17, No. 4, November 2009.

Shaffer, Mark L. "Minimum Population Sizes for Species Conservation." *BioScience,* Vol. 31, No. 2, February 1, 1981. doi:10.2307/1308256.

Simpson, George Gaylord. *Tempo and Mode in Evolution.* New York: Columbia University Press, 1944.

Smith, Thomas B., Michael T. Kinnison, Sharon Y. Strauss, Trevon L. Fuller, and Scott P. Carroll. "Prescriptive Evolution to Conserve and Manage Biodiversity." *Annual Review of Ecology, Evolution, and Systematics*, Vol. 45, No. 1, November 23, 2014. doi:10.1146/annurev-ecolsys-120213-091747.

Sodikoff, Genese Marie, ed. *The Anthropology of Extinction: Essays on Culture and Species Death.* Bloomington: Indiana University Press, 2011.

Soulé, Michael E. "Thresholds for Survival: Maintaining Fitness and Evolutionary Potential." *Conservation Biology: An Evolutionary-Ecological Perspective*, Vol. 111, 1980.

———. "What Is Conservation Biology?" *BioScience*, Vol. 35, No. 11, December 1985. doi: 10.2307/1310054.

———. "The 'New Conservation.'" *Conservation Biology*, Vol. 27, No. 5, October 2013. doi:10.111/cobi.12147.

Soulé, Michael E, and Bruce A. Wilcox, eds. "Conservation Biology. An Evolutionary-Ecological Perspective." Sunderland: Sinauer Associates, 1980.

Steiner, Cynthia. "Looking at Alala Genome." San Diego Zoo. *Wildlife Field Notes: Firsthand Experiences With Saving Endangered Species*, December 6, 2013. http://blog.sandiego zooglobal.org/2013/12/06/looking-at-alala-genomes/.

Stelkens, Rike B., Michael A. Brockhurst, Gregory D. D. Hurst, and Duncan Greig. "Hybridization Facilitates Evolutionary Rescue." *Evolutionary Applications*, Vol. 7, Issue. 10, September 1, 2014. doi:10.1111/eva.12214.

Stewart, G., K. Mengersen, G. M. Mace, J. A. McNeely, J. Pitchforth, and B. Collen. "To Fund or Not to Fund: Using Bayesian Networks to Make Decisions about Conserving Our World's Endangered Species." *Chance: Magazine of the American Statistical Association*, 2013.

Stockwell, Craig A., Jeffrey S. Heilveil, and Kevin Purcell. "Estimating Divergence Time for Two Evolutionarily Significant Units of a Protected Fish Species." *Conservation Genetics*, Vol. 14, No. 1, February 2013. doi:10.1007/s10592-013-0447-1.

Stockwell, Craig A., Andrew P. Hendry, and Michael T. Kinnison. "Contemporary Evolution Meets Conservation Biology." *Trends in Ecology & Evolution*, Vol. 18, No. 2, 2003.

Stockwell, Craig A., and Paul L. Leberg. "Ecological Genetics and the Translocation of Native Fishes: Emerging Experimental Approaches." *Western North American Naturalist*, Vol. 62, No. 1, 2002.

Stockwell, Craig A., Margaret Mulvey, and Adam G. Jones. "Genetic Evidence for Two Evolutionarily Significant Units of White Sands Pupfish." *Animal Conservation*, Vol. 1, No. 3, August 1, 1998. doi:10.1111/j.1469-1795.1998.tb00031.x.

Stockwell, Craig A., and Stephen C. Weeks. "Translocations and Rapid Evolutionary Responses in Recently Established Populations of Western Mosquitofish (Gambusia Affinis)." *Animal Conservation*, Vol. 2, No. 02 (1999): 103–10.

"Surviving Climate Change May Be Genetic According to Trent University Research." *Trent University*, April 25, 2012. http://www.trentu.ca/newsevents/newsDetail.php?newsID =2485.

Swaisgood, Ronald R., and James K. Sheppard. "The Culture of Conservation Biologists: Show Me the Hope!" *BioScience*, Vol. 60, No. 8, September 1, 2010. doi:10.1525/bio.2010.60.8.8.

Sylvan (formerly Routley), Richard. "Is There a Need for a New, an Environmental, Ethic?" In *XVth World Congress of Philosophy*, No. 1.Varna, Bulgaria: Sofia Press, 1973.

Thatcher, Cindy A., Frank T. van Manen, and J. D. Clark. "An Assessment of Habitat North of the Caloosahatchee River for Florida Panthers." University of Tennessee and US Geological Survey, Knoxville, TN. Final Report to US Fish and Wildlife Service, Vero Beach, FL, 2006.

Thomas, Nilsen. "No Ice—No Cubs." *Barents Observer*, June 27, 2012. http://barentsobserver .com/en/nature/no-ice-no-cubs-27-06.

Tonnesen, Gail. "Mitochondrial DNA Haplogroup U5: Description of mtDNA Haplogroup U5." *Family Tree DNA*, July 18, 2014. https://www.familytreedna.com/public/u5b/default .aspx?section=results.

"TRAFFIC—Wildlife Trade News—Pioneering Research Reveals New Insights into the Consumers behind Rhino Poaching." *Traffic: The Wildlife Trade Monitoring Network*, September

17,   2013.   http://www.traffic.org/home/2013/9/17/pioneering-research-reveals-new
-insights-into-the-consumers.html.

Tuck, Robert A., and Robert Grenier. "A 16th-Century Basque Whaling Station in Labrador."
Scientific American, Vol. 245, No. 5, 1981.

Umbreit, Andreas Dr. *Svalbard: Spitzbergen, Jan Mayen, Frank Josef Land*. Fifth edition. Buck-
inghamshire, UK: Bradt Travel Guides, 2013.

U.S. Fish and Wildlife Service. *Final Environmental Assessment: Genetic Restoration of the Florida Panther*. Gainesville, Florida, December 20, 1994.

U.S. Seal and the Workshop Participants. *Genetic Management Strategy and Population Viability
of the Florida Panther (Felis Concolor Coryi)*. National Zoological Park, Washington, DC
and White Oak Plantation Conservation Center, Yulee, Florida: Captive Breeding Specialist Group SSC/IUCN, May 30, 1991.

Van de Lavoir, Marie-Cecile, Ellen J. Collarini, Philip A. Leighton, Jeffrey Fesler, Daniel R. Lu,
William D. Harriman, T. S. Thiyagasundaram, and Robert J. Etches. "Interspecific Germ-
line Transmission of Cultured Primordial Germ Cells." Edited by Osman El-Maarri. *PLoS
ONE*, Vol. 7, No. 5, e35664, May 21, 2012. doi:10.1371/journal.pone.0035664.

Vander Wal, E., D. Garant, M. Festa-Bianchet, and F. Pelletier. "Evolutionary Rescue in Vertebrates: Evidence, Applications and Uncertainty." *Philosophical Transactions of the
Royal Society B: Biological Sciences*, Vol. 368, No. 1610, December 3, 2012. doi:10.1098/
rstb.2012.0090.

Van Dooren, T. *Flight Ways: Life and Loss at the Edge of Extinction*. New York: Columbia University Press, 2014.

———. "Authentic Crows: Identity, Captivity and Emergent Forms of Life." *Theory, Culture and
Society*, forthcoming.

———. "Banking the Forest: Loss, Hope and Care in Hawaiian Conservation." In *Defrost:
New Perspectives on Temperature, Time, and Survival*, edited by Joanna Radin and Emma
Kowal, forthcoming.

Walters, Mark Jerome. *Seeking the Sacred Raven: Politics and Extinction on a Hawaiian Island*.
Washington, DC: Island Press, 2006.

Walton, Murray T. "Rancher Use of Livestock Protection Collars in Texas." In *Proceedings of the
Fourteenth Vertebrate Pest Conference 1990*, 80, 1990.

Weidensaul, Scott. *The Ghost with Trembling Wings: Science, Wishful Thinking and the Search for
Lost Species*. New York: North Point Press, 2003.

Weldon, Ché. "Chytridiomycosis, an Emerging Infectious Disease of Amphibians in South
Africa." Thesis, North-West University, 2005. http://dspace.nwu.ac.za/handle/10394/860.

Weldon, Ché, Louis H. du Preez, Alex D. Hyatt, Reinhold Muller, and Rick Speare. "Origin of
the Amphibian Chytrid Fungus." *Emerging Infectious Diseases*, Vol. 10, No. 12, December
2004. doi:10.3201/eid1012.030804.

West, Paige. *Conservation Is Our Government Now: The Politics of Ecology in Papua New Guinea*.
Durham: Duke University Press Books, 2006.

White, Lynn Jr. "The Historical Roots of Our Ecological Crisis." *Environmental Ethics: Readings
in Theory and Application*, Belmont: Wadsworth Company, 1998.

Wiley, E. O. "The Evolutionary Species Concept Reconsidered." *Systematic Biology*, Vol. 27, No.
1, March 1, 1978. doi:10.2307/2412809.

Williams, Nigel. "Fears Grow for Amphibians." *Current Biology*, Vol. 14, No. 23, December 14,
2004. doi:10.1016/j.cub.2004.11.016.

Wynn, Thomas, and Frederick L. Coolidge. *How To Think Like a Neandertal*. New York: Oxford
University Press, 2011.

Young, S. P., and E. A. Goldman. "Puma, Mysterious American Cat: Part I: History, Life Habits,
Economic Status, and Control." American Wilderness Institution, Washington DC, 1946.

Zippel, Kevin, Kevin Johnson, Ron Gagliardo, Richard Gibson, Michael McFadden, Robert
Browne, Carlos Martinez, and Elizabeth Townsend. "The Amphibian Ark: A Global Com-
munity for Ex Situ Conservation of Amphibians." *Herpetological Conservation and Biology*,
Vol. 6, No. 3, December 2011.

INDEX